PAPER 1.
Product Law

ALL BASIC PHYSICAL ACTIONS (EFFECTS) ARE **PRODUCT** OF TWO TERMS, AND ALL PHYSICAL INTERACTIONS ARE **PRODUCTS** OF MULTIPLE TERMS.

If an entity is present in space, it has a physical action (its effect) in the space around. This physical action at a distance d from the entity,

1. is proportional to the quantity of this entity

2. and is inversely proportional to the distance d (as d increases, the strength of this physical action decreases).

This physical action is the product of two physical quantities (quantity of the entity) and $(1/d)$. If this entity is mass, its physical action (effect) at a distance d is m x $1/d$ (product of mass m and inverse of distance d). Two physical actions of same type create an interaction.

To feel this physical effect, there should be another entity of same type in the neighborhood. The presence of another entity of same type creates an interaction between these two physical actions. This interaction will be products of physical quantities.

If this entity is mass, another entity of same type is mass. If this entity is electric charge, another entity of same type is electric charge.

Applications of this product law are given in chapter 3 to chapter 4.2.

TABLE OF CONTENTS

PAPER NO.	DESCRIPTION	PAGE
	This table	1-2
	Preface	3
	Procedure followed in this book in flowchart format	4
	This page left blank	5
1	Product Law	6
2	Explanations of physical equations	7
3	Gravitational force between two masses	11
4.1	Electrostatic force between a positive and a negative charge	13
4.2	Electrostatic force between similar charges	15
5.1	Nuclear mass defect – Nuclear, proton and neutron shocks	18
5.2	Nuclear Force Also see Paper 27, material added to this in October 2022	24
6	Photon-Electron interaction	31
7	Multiple Photon Absorption – Photo electric effect	33
8	A wave produced by an oscillating or a vibrating object	35
9	Signed Physical Equations Constants	41
10	Mass effect and charge effects	43
11	Accelerating particles radiate energy	44
12	Newton's laws of motion	46
13	Speed of light and the energy per unit mass and particles as circulating energy packets	48
14	Gravitational, electrostatic and nuclear forces	50
15	Latent heat of fusion and vaporization	51
16	Measured and effective values in physical equations and the conversion factor	53
17	Newton's second law of motion derived from his first law	56
18	General equations of motion and the equation of the conservation of mechanical energy in a free fall	60
19	General equations of motion and the equation of the conservation of mechanical energy in a pendulum motion.	72
20	Measure the speed of Earth around sun	75
21	A stationary pendulum is a substitute for Foucault Pendulum	81
22	General equations of motion are the sum of the parts of initial velocity and the acceleration.	83
23	Space, time, and distance relationship	89

TABLE OF CONTENTS (continued)

PAPER NO.	DESCRIPTION	PAGE
24	The two states of matter – the state of rest and the state of uniform motion in a straight line	92
25	Modified Statement of Newton's third law of motion	93
26	Strong force of repulsion in superconductors	95
27	Material added to Paper 5.2	97
28	Action, Reaction and Centrifugal Force	98
29	Reference frames and first postulate of special theory of relativity	101
30	Entangled Photons	104
31	Rules of the basic laws of physics	105
32	Floating or sinking body in a fluid	106
33	Particle and Space Signals	107

PREFACE

Fundamental approach in this book is asking meaningful questions in physical observations and try to get sensible answers. Sensible analyses of various physical observations are performed in this book. This approach leads to new explanations of some observations and new predictions. Basic laws of nature are not a set of complicated mathematics. These laws are digestible to common sense. Human sense is common sense.

Some unconventional procedures and research methods are used to search for answers/explanations to questions in the field of physics. Few of my findings and explanations are presented in this book.

General approach to physical observations in this book is the following. Reader's attention is drawn to various physical observations and walks them through an in-depth analysis. Then proceeds to raise thought-provoking questions, some never asked, and some never answered (see the procedure followed in this book in flowchart format on page 3). These questions lead to some sensible answers. In some cases, the answers are based on existing laws; in others, the answers lead to new logical predictions. Despite the intricacies of the topics, attempts are made to make the physical discussions, questions raised and their discussions, intelligible to anyone used to making commonsense observations—in many cases, no background in physics required.

General approach to different physical phenomena followed in this book is given in a flowchart format in this book. Some new concepts introduced are listed after this flowchart.

Explanations I have given here may not reasonably be the final answer. Further questions can always be asked, and deeper insights to the secrets of nature can be obtained.

Jacob V. Kainnady, M.Sc., Ph.D.

PHYSICS The Science of Common Sense (Third Edition)
Copyright © 2020 by Jacob V. Kainnady.
All rights reserved.

For information about the material in this document or
Permission to use material from this document,
Contact Jacob V. Kainnady by Email - profkainnady@gmail.com

PROCEDURE FOLLOWED IN THIS BOOK IN FLOW CHART FORMAT

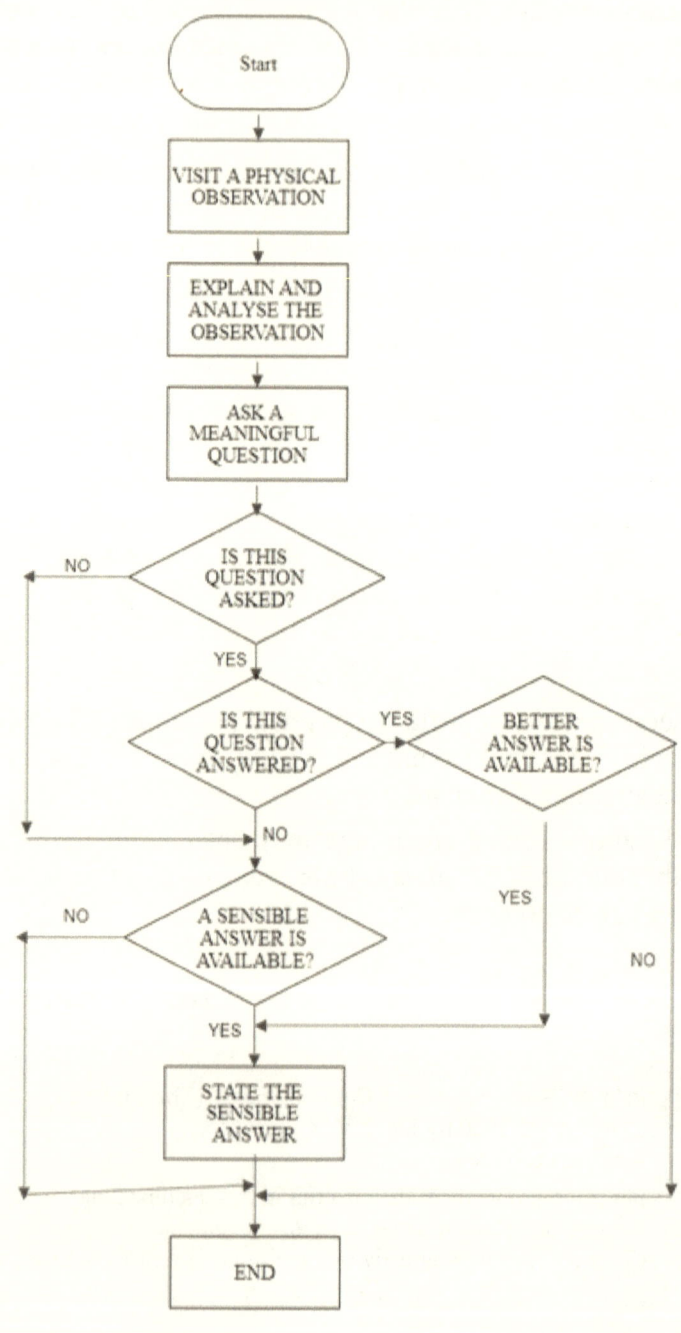

PAPER 2.
EXPLANATIONS OF PHYSICAL EQUATIONS

Every physical equation deserves a meaningful explanation.

Every physical equation should be formatted properly so it can be understood very easily. Formatted equations should make the more sense. That is, format it into a common-sense equation+. Here some physical equations are formatted to give meaningful explanations.

1. Consider the speed of a wave,

v (distance travels in unit time) = λ (wavelength) x ν (frequency)
Distance travels in unit time = (Number of waves in unit time) x
(Length of one wave) ………... (1)

This common-sense equation makes more sense.

2. Mass-energy equation is,

$E = mc^2$

Energy = Mass x (speed of light)^2

This equation can be written as,

Energy of mass m = Mass (m) x Energy per unit mass (c^2) …… (2)

This common-sense equation makes more sense.

3. Considering the quantum theory, equation for the energy (E) of a photon is

$E = h$ (Planck's constant) x ν (frequency)

$= \nu \times h$

This equation can be written as,

Energy of a photon with frequency ν
= Number of waves (cycles) per second (ν) x h
= Number of waves (cycles) per second (ν) x
Energy in one wave (cycle) …… (3)

This common-sense equation makes more sense.

4. Gravitational force *F* between 2 masses

M1 and M2 separated by a distance x is

$$F = k \frac{M1 \times M2}{x^2} \text{ here is a constant}$$

$$F = k \frac{M1}{x^2} M2$$

Here $k \frac{M1}{x^2}$ is the force of mass M1 on a unit mass at a distance x from M1.

So, the above equation can be written as,

Force of a mass M1 on a mass M2 placed at a distance x from M1
= (Force by M1 on a unit mass at a distance x) x
(mass M2 at a distance x from M1). ......
= (mass M2 at a distance x from M1) x
(Force by M1 on a unit mass at a distance x from M1)....(4)
OR
= (mass M1 at a distance x from M2) x
(Force by M2 on a unit mass at a distance x from M2)....(4)

This common-sense equation makes more sense.

Electrostatic force between 2 charges can also be given a meaningful explanation like this.

5. Newton's second law is

Force F on a mass $m = m\,a,$ here a is the acceleration

Force F on a mass $m = Mass\,(m) \times acceleration(a)$

This equation can be written as,

Force F on a mass m = Mass (m) x Force on a unit mass ...... (5)

This common-sense equation makes more sense.

6. Centrifugal Force. An object of mass m traveling with speed of v along a circle of radius r experiences an outward force called centrifugal force.

$$This\ centrifugal\ force,\ F = m \frac{v^2}{r}$$

That is, $F = m\,a,$ where a is the acceleration

This equation can be written as,

Centrifugal force *on mass m = Mass (m)* ×

Centrifugal force on a unit mass .. (6)

This common-sense equation makes more sense.

7. Gravitational potential energy. Gravitational potential energy of a mass m near earth is

Potential Energy = mgh

= m (gh)

This equation can be written as,

P. E. of a mass m = mass (m) x gravitational Potential energy of unit mass ..(7)

This common-sense equation makes more sense.

Table1 lists regular formats of various physical equations and their rephrased formats that make more sense.

TABLE 1 – REGULAR AND REPHRAISED PHYSICAL EQUATIONS

NO.	REGULAR EQUATION	REFORMATTED EQUATION THAT MAKES THE MORE SENSE
1	v (distance travels in unit time) = λ (wave length) x ν (frequency) $v = \lambda \nu$	v (distance travels in unit time) = Number of waves in unit time x Length of one wave
2	Mass-energy equation $E = mc^2$	Energy = Mass (m) x Energy per unit mass
3	Equation for the energy (E) of a photon. E = h (Planck's constant) x ν (frequency)	E = Frequency (ν) x Energy per unit frequency
4	Gravitational force between 2 masses separated by a distance x $F = k \frac{M1}{x^2} M2$	Gravitational force between 2 masses M1 and M2 = (Force of M1 on a unit mass at a distance x) x (mass M2 at a distance x).
5	Newton's second law. Force F on mass m = m a	Force F on mass m = Mass (m) x Force on a unit mass
6	Centrifugal Force. F on mass $m = m \frac{v^2}{r}$	$F = m \frac{v^2}{r}$ Centrifugal force on mass m = Mass (m) x Centri. force on a unit mass
7	Gravitational potential energy of mass m near earth. Potential Energy = mgh	P. E. of mass m = Mass (m) x P. E. of unit mass

⁺ A simple example of a *common-sense equation* is the following.

Consider N number of balls of equal masses (x) in a container.
Total mass of the balls in the container = Number of balls (N)
                                                        Mass of one ball (x)

In general,

> Measure of a physical quantity = Number of units x Unit measure

PAPER 3.
GRAVITATIONAL FORCE BETWEEN TWO MASSES

It is good to know the logical evolution (a little bit of history of scientific discoveries) that led Newton to his universal law of gravitation. Nicolaus Copernicus, born in Poland in 1473, known as father of modern astronomy developed the heliocentric model of Solar System. Galileo Galilei, born in Italy in 1564, developed modern optics. using telescopes and observational astronomy. Johannes Kepler, born in Germany in 1571, developed his famous three laws of Planetary Motion describing then known planets (Mercury, Venus, Earth, Mars, Jupiter, and Saturn) around sun. Working from these, Sir Isaac Newton developed his Universal Law of Gravitation. Newton had the modesty to express the debt he owed to the discoveries of Galileo and Keppler. In his letter written in 1676 to English scientist he wrote "If I have been further, it was by standing on the shoulders of giants"

Isaac Newton introduced the law of gravitation in 1687 which explained motions of the planets and their moons.

According to Newton's law of gravitation, the force (attraction) between 2 masses $M1$ and $M2$ separated by a distance x is -

$F = k \frac{M1 \times M2}{x^2}$ where k is a constant. This is the universal gravitational constant G.

This law is verified. Consider whether this law can be further explained.

Two questions about this equation are -
1. why the product of the masses come into this equation?
2. Why is this force inversely proportional to the square of the distance between the masses?

Consider a mass $M1$ at any point in space. The mass $M1$ exhibits its presence in its surrounding space.

This mass has a physical action (effect), a mass effect $M1e$, at a point A distant x away from it. $M1e$ is the physical action of mass $M1$ *at point A*. This effect is neither attraction nor repulsion and can be measured only by the interaction on a similar entity (another mass). The mass $M1$ sees point A at a distance x and not at x^2 from it. That is, the effect $M1e$ at point A depends on the distance x and not on x^2.

$M1e$ depends on

1. The distance x.
   As the distance x increases, $M1e$ decreases.
2. The quantity of the mass $M1$.
   As the quantity of the mass $M1$ increases, $M1e$ increases.

That is, $M1e = k_1 \frac{M1}{x}$ here k_1 is a constant.

Now place another mass $M2$ at point A. This mass has its effect (physical action) $M2e$ on mass $M1$. $M2e$ is a similar effect as $M1e$.

That is, $M2e = k_2 \frac{M2}{x}$ here k_2 is a constant.

Here there are 2 masses,

1. Mass $M1$ under the physical action (mass effect) $M2e$ of mass $M2$ and
2. Mass $M2$ under the physical action (mass effect) $M1e$ of mass $M1$.

These masses are separated by a distance x.

These two physical actions (effects), $M1e$ and $M2e$ interact with each other. The result of this interaction is the product (see Paper1 in this book) of the two ($M1e$ and $M2e$) and is an attractive force between the masses.

Thus, each mass attracts other with a force

$$F = M1e \times M2e = k \frac{M1 \times M2}{x^2}$$ here k is a constant. This is the universal gravitational constant G.

This is Newton's law of gravitational force between the 2 masses.

Notice the mass effects of both masses are the product of two physical quantities. Also, the interaction between the two mass effects is also the product of two physical quantities.

PAPER 4.1

ELECTROSTATIC FORCE BETWEEN A POSITIVE AND A NEGATIVE CHARGE

Force (attraction) between a positive charge P and a negative charge E separated by a distance x is

$$F = k \frac{P \times E}{x^2}$$ where k is a constant (Coulomb's constant).

This law is verified. Consider whether this law can be further explained.

Two questions about this equation are

1. why the product of the charges come in this equation?
2. why this force is inversely proportional to the square of the distance between the charges?

Consider a positive charge P at any point in space. The charge P exhibits its presence in its surrounding space.

This positive charge has a physical action (effect), a charge effect Pe, at a point A distant x away from it. Pe is the physical action of charge P at point A. This effect is neither attraction nor repulsion and can be measured only by the interaction on a similar entity (another charge – positive/negative). This positive charge P sees the point A at a distance x and not at x^2 from it. That is, this effect Pe at point A depends on the distance x and not on x^2.

Pe depends on

1. The distance x.
   As the distance x increases, Pe decreases.
2. The charge P.
   As the amount of charge P increases, Pe increases.

That is, $P_e = k_1 \frac{P}{x}$ here k_1 is a constant.

Now place a negative charge E at point A. This negative charge has its physical action (effect) Ee on the positive charge P. Ee is a similar effect as Pe.

That is, $Ee = k_2 \frac{E}{x}$ here k_2 is a constant.

Here there are 2 charges.

1. A positive charge P under a charge effect Ee of a negative charge E and
2. A negative charge E under a charge effect Pe of a positive charge P.

These charges are separated by a distance x.

These two physical actions (effects) interact with each other. The result of this interaction is the product of these physical actions (effects), Pe and Ee according to the product law (see Paper1 in this book). The resulting interaction turns out to be an attractive force on each charge towards the other charge.

That is, the interaction (attraction) between the 2 opposite charges P and E separated by a distance x is

$$F = Pe \times Ee = k \frac{P \times E}{x^2}$$ here k is a constant (Coulomb's constant).

This is the electrostatic force of attraction between 2 opposite charges.

Notice the charge effects of both charges are the product of two physical quantities. Also, the interaction between the two charge effects is also the product of two physical quantities

PAPER 4.2.
ELECTROSTATIC FORCE BETWEEN SIMILAR CHARGES

1. Consider force between two positive charges.

Force (repulsion) between two positive charges $P1$ and $P2$ separated by a distance x is

$$F = k \frac{P1 \times P2}{x^2}$$ where k is a constant (Coulomb's constant).

This law is verified. Consider whether this law can be further explained.

Two questions about this equation are

1. why the product of the charges come in this equation?
2. why this force is inversely proportional to the square of the distance between the charges?

Consider a positive charge $P1$ at any point in space. The charge $P1$ exhibits its presence in its surrounding space.

This positive charge has a physical action (effect), a charge effect $P1e$, at a point A distant x away from it. $P1e$ is the physical action of charge $P1$ at point A. This effect is neither attraction nor repulsion and can be measured only by the interaction on a similar entity (another charge – positive/negative). This positive charge $P1$ sees the point A at a distance x and not at x^2 from it. That is, this effect $P1e$ at point A depends on the distance x and not on x^2.

$P1e$ depends on

1. The distance x.
   As the distance x increases, $P1e$ decreases.

2. The charge $P1$.
   As the amount of charge $P1$ increases, $P1e$ increases.

That is, $P1e = k_1 \frac{P1}{x}$ here k_1 is a constant.

Now place another positive charge $P2$ at point A. This positive charge has its physical action (effect) $P2e$ on the positive charge $P1$. $P2e$ is a similar effect as $P1e$.

That is, $P2e = k_2 \dfrac{P2}{x}$ here k_2 is a constant.

Here there are two charges.
1. A positive charge $P1$ under a charge effect $P2e$ of a positive charge $P2$ and
2. Another positive charge $P2$ under a charge effect $P1e$ of a positive charge $P1$

These charges are separated by a distance x.

These two physical actions (effects) interact with each other. The result of this interaction is the product of these physical actions (effects), $P1e$ and $P2e$ according to the product law (see Paper1 in this book). The resulting interaction turns out to be a repulsive force on each charge away from the other charge.

That is, the interaction (repulsion) between the 2 positive charges $P1$ and $P2$ separated by a distance x is

$$F = P1_e \times P2_e = k \dfrac{P1 \times P2}{x^2}$$ here k is a constant (Coulomb's constant).

This is the electrostatic force of repulsion between 2 positive charges.

2. Force between two negative charges.

The force (repulsion) between two negative charges also can be explained like this.

PAPER 5.1

NUCLEAR MASS DEFECT – NUCLEAR, PROTON AND NEUTRON SHOCKS

During the fusion of 2 particles, it is observed that energy is released. The energy released is equivalent to the mass defect. Mass defect is the difference between the mass of the nucleus and the sum of the masses of the independent nucleons.

Consider the mass of a particle (proton/neutron) outside an atom (m_{pout}) and mass of a proton inside a nucleus (m_{pin}).

It is found that $m_{pout} > m_{pin}$

$$m_{pout} = \Delta m + m_{pin}$$

Here, Δm is the mass defect of the particle.

Energy equivalent to Δm is released during the binding of the nucleus. **Since this energy is lost, it cannot be the binding energy of the nucleons. The nucleons are sticking together by the strong nuclear forces of attraction.** This attractive force is explained in Paper 5.2.

Here, Δm is the mass defect of the particle. Mass defect is applicable for nucleus and all nucleons (protons and neutrons).

Average mass defect per nucleon is calculated as follows.

Consider copper-63 atom. Its nucleus has 29 protons and 34 neutrons.

The mass of a proton (outside the nucleus) $m_p = 1.00728$ amu

The mass of a neutron (outside the nucleus) $m_n = 1.00867$ amu

Total mass of 29 protons + 34 neutrons = (29 × 1.00728 amu)
+ (34 × 1.00867 amu) = 63.50590 amu

Measured mass of copper-63 nucleus is 62.91367 amu

Total nuclear mass defect = 63.50590 amu – 62.91367 amu
= 0.59223 amu

Average mass defect per nucleon of copper-63 nucleus
= 0.59223 amu ÷ 63 = 0.0094 amu/nucleon

Using $E = mc^2$, energy equivalent of average nucleon mass defect can be calculated as 8.758 Mev

What causes this mass defect (release of energy)?

1. Consider a proton approaching a nucleus.

The electrostatic force of repulsion between the positively charged nucleus with a charge Q and the proton with a charge q separated by a distance x is according to Coulomb's law.

$$F = k \frac{Qq}{x^2} \quad \text{here k is a constant}$$

When the distance x becomes smaller, the electrostatic force of repulsion increases. This repulsive force is very large just outside the nuclear range.

If the proton is brought still closer to the nucleus, (reduce x further) they are in the nuclear range and the strong nuclear force of attraction is dominant (**FIGURE-1**).

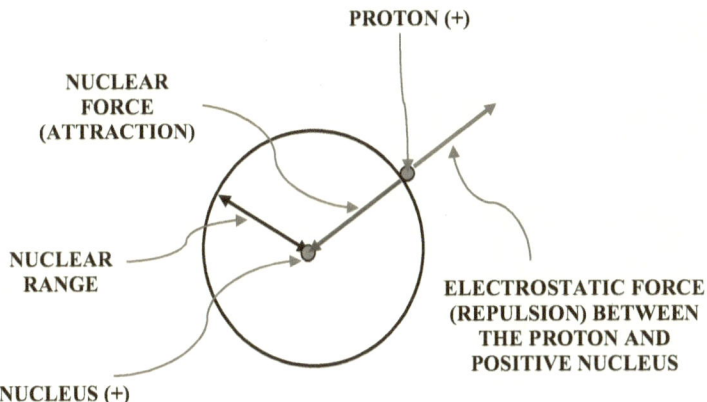

FIGURE-1

Thus, just outside the nuclear range, the positive nucleus and the proton experience a strong electrostatic force of repulsion and just inside the nuclear range these two objects experience a strong nuclear force of attraction.

This reversal of the direction of very strong forces (in a short range of distance) result in the **nuclear and the particle (proton) shocks**. That is the nucleus and the proton experience accelerations.

An accelerated charge radiates energy according to Larmor formula. According to Larmor, power radiated by an accelerating charge

$$P = \frac{2}{3}\frac{q^2 a^2}{c^3}$$

here q is the charge, a is the acceleration and c is the speed of light.

So nuclear and the proton shocks (acceleration of charged particles) cause some of the mass (of the nucleus and of the proton) to convert to energy. This combined mass (of nuclear and proton defects) converted to energy is the total nuclear mass defect.

Greater the nuclear charge, greater is the nuclear acceleration and the resulting nuclear mass defect. Greater the nuclear mass, lesser is the nuclear acceleration and the resulting nuclear mass defect.

2. Consider a neutron approaching a nucleus.

Outside the nuclear range the force between the nucleus of mass M and the neutron of mass m separated by a distance x is the weak gravitational force of attraction.

$$F = G \frac{M m}{x^2} \quad \text{here G is a constant}$$

When a neutron is brought from outside the nuclear range to inside the nuclear range, this gravitational force of attraction is overcome by strong nuclear force of attraction which causes a **neutron shock (acceleration)** - **FIGURE-2**. A neutron in the nuclear range already bound with the nucleus. So, a shock of this neutron causes a shock of the nucleus also. That is the nucleus and the neutron experience accelerations.

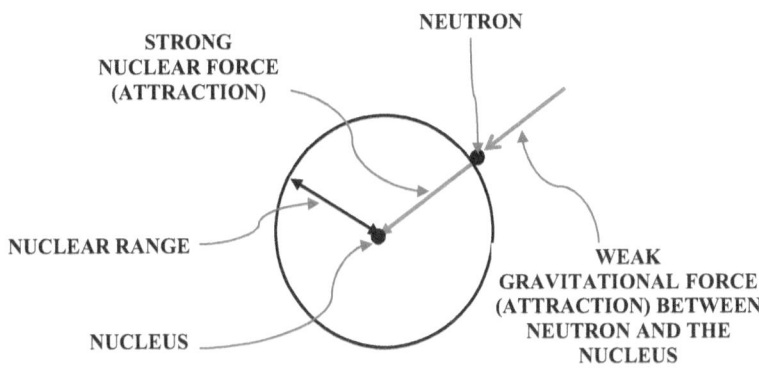

FIGURE-2

Acceleration of charged particle radiates energy according to Larmor formula. An accelerated charge radiates energy according to Larmor formula. According to Larmor, power radiated by an accelerating charge

$$P = \frac{2}{3} \frac{q^2 a^2}{c^3}$$

here q is the charge, a is the acceleration and c is the speed of light.

That is some nuclear mass is converted to energy due to nuclear acceleration. This mass converted to energy is the mass defect of the nucleus. According to paper 11, in this book, accelerated neutron also radiates energy and so has a mass defect. This combined mass (of nucleus and neutron) converted to energy is the total nuclear mass defect.

As in the case of the proton entry to the nucleus, greater the nuclear charge, greater is the nuclear acceleration and the resulting nuclear mass defect. Greater the nuclear mass, lesser is the nuclear acceleration and the resulting nuclear mass defect.

3. Summary and conclusion.

Mass defect per nucleon varies with the number of nucleons in a nucleus.

- 3.1. Nuclear mass defect is due to nuclear acceleration.
- 3.2. Particle mass defect is due to particle acceleration.
- 3.3. Increase in the nuclear charge increases the nuclear acceleration (by Larmor) resulting in the increase in the nuclear mass defect. Increase in the nuclear mass decreases the nuclear acceleration by Newton's second law which results in reduced nuclear mass defect.
- 3.4. Repulsive electrostatic force between a nucleus and a proton is stronger than the attractive gravitational force between a nucleus and a neutron. So, proton will have greater acceleration and greater mass defect compared to the acceleration and the mass defect of a neutron.
- 3.5. Total mass defect of the nucleus is the sum of these two mass defects. That is,
$$Total\ mass\ defect = Mass\ defect\ of\ the\ nucleus + Mass\ defect\ of\ the\ incoming\ particle.$$
- 3.6. Mass defect per nucleon $= \dfrac{Total\ mass\ defect}{Total\ number\ of\ nucleons}$

  $= \dfrac{Mass\ defect\ of\ the\ nucleus + Mass\ defect\ of\ the\ incoming\ particle}{Number\ of\ nucleons\ including\ the\ incoming\ particle}$
- 3.7. It seems the following are the roles of the charge and the mass of a nucleus in the nuclear acceleration.
    - 3.7.1. For nuclei of smaller masses, the nuclear charge dominates to produce more acceleration compared to the nuclear mass which reduces the acceleration. This results in considerable nuclear mass defect.
    - 3.7.2. After a nucleus obtains certain mass, the nuclear mass dominates to reduce the acceleration compared to the nuclear charge which increases the acceleration. This results in lesser nuclear mass defect.

As this reduction in nuclear mass defect continues, the nuclear mass defect becomes exceedingly small compared to the mass defect of the incoming particle.

3.8. Consider equation in 3.6 as each particle is added to the nucleus –
   3.8.1. The denominator of equation in 3.6 increases by 1.
   3.8.2. Starting from the lightest nucleus (hydrogen) going to higher nuclear masses, the numerator of equation in 3.6 increases.
   3.8.3. After a nucleus obtains certain mass, the increase of numerator in equation 3.6 becomes smaller and reaches a value almost equals to the mass defect of the particle being added.

3.9. **So as each particle is added to a nucleus, after a certain nuclear mass, the mass defect per nucleon decreases. This makes the nucleus less stable.**

3.10. **Mass defects can be calculated based on the explanation given here with the following and related known values**.
   1. charge of the nucleus and a proton
   3. masses of the nucleus. proton and neutron
   4. electrostatic force between the nucleus and a proton
   5. gravitational force between the nucleus and a neutron
   6. nuclear force
   7. nuclear force range

PAPER 5.2
NUCLEAR FORCE

During the fusion of 2 particles, it is observed that energy is released. The energy released is equivalent to the mass defect. Mass defect is the difference between the mass of the nucleus and the sum of the masses of the independent nucleons. This energy is released. This lost energy cannot be the binding energy of the nucleons. The nucleons are sticking together by strong nuclear forces of attraction.

What is this nuclear force (force of attraction) due to?

Nature assigns specific physical properties (mass, charge etc.) to various elementary particles. These assigned values make the particles stable. For example, all protons have same mass and charge. If an elementary particle gets a physical property more than its nature assigned value, it is less stable and may disintegrate. **If a particle has a physical property less than the nature assigned value, it may combine/share its deficiency with another particle so both particles together will be stable. This is like sharing of valence electrons when 2 atoms are chemically combined to make the valence orbits of both atoms complete and both atoms together are stable.**

Mass defect is applicable for nucleus and nucleons (see Paper 5.1). Mass defects for nuclei and for different nucleons may be different.

1. **The nuclear force which hold the nucleons together is not the energy released during fusion. But it is due to the mass defects of the nucleus and nucleons, which are being shared among themselves and thus glue the nucleons together**.

   For the following descriptions consider Figure-1 in this paper.
   1. Start with a nucleus of 1 particle A (a proton).
   2. Particle B joins the nucleus.
      Now ΔmA is the mass defect of nucleus (made of particle A) which is shared from B.

      ΔmB is the mass defect of particle B which is shared from A.

   These 2 nucleons are highly stable as long as they stay together sharing their mass deficiencies. That is these nucleons stick together with great force. This is the nuclear force.

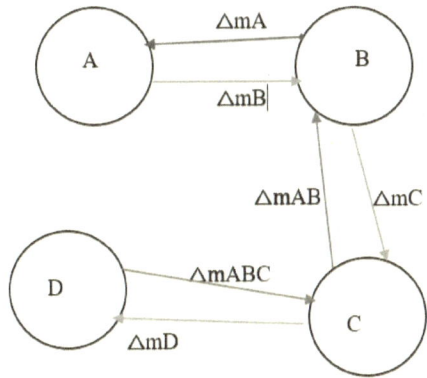

FIGURE-1

3. Particle C joins the nucleus consists of A and B.

Now ΔmAB is the mass defect of nucleus (made of A and B) which is shared from C.

ΔmC is the mass defect of particle C which is shared from the nucleus (made of A and B).

Then these 3 nucleons together will be sticking together (fused together) and will be stable as one piece.

As more nucleons are present, the same process continues.

This sharing of masses between nucleons is the cause of the nuclear binding force. This force spreads to its immediate neighborhood (very short ranged).

The greater the amount of mass defect shared between 2 nucleons, greater is the nuclear binding force between these 2 nucleons and lesser the amount of mass defect shared between 2 nucleons, lesser is the nuclear binding force between these 2 nucleons.

2. When the number of nucleons increases (nucleus becomes bigger), after a certain mass the nucleus becomes more unstable.

2.1. When the number of nucleons increases, the nuclear mass increases.
2.2. After a certain nuclear mass, the mass defect per nucleon decreases – see 3.9 in Paper 5.1.
2.3. Since mass defect is shared between the nucleons, after a certain nuclear mass, the mass defect shared between the nucleons decreases.
2.4. This decreases the average nuclear binding force between nucleons.
2.5. Reduction of the average nuclear binding force makes the nucleus less stable.
2.6. That is, as the nucleus becomes heavier, after a certain mass, it becomes less stable.

3. The experimental graph (Graph-1) can be explained as follows.

3.1. For lighter nuclei, there is relatively considerable nuclear mass defect in addition to the particle mass defect – see 3.7.1 in Paper 5.1.
3.2. That means for lighter nuclei, the sharing of the mass defect per nucleon is relatively considerable.
3.3. As the nucleus becomes heavier, after a certain mass, the mass defect per nucleon decreases – see 3.9 in Paper 5.1.
3.4. That means as the nucleus becomes heavier, after a certain mass, the sharing of the mass defect per nucleon decreases.
3.5. That is as the mass of the nucleus increases, after a certain mass, the shared mass defect between the nucleons decreases.
3.6. This results in the reduction of average binding energy per nucleon for heavier nuclei.
3.7. **These explanations agree with the experimental curve –** *Average binding energy per nucleon vs Number of nucleons in nucleus graph* **– see Graph-1 in this paper.**

The nucleons take care of their mass defects themselves. Sharing is caring. Nucleons works on the principle that sharing they stand divide they fall!

Average binding energy per nucleon vs Number of nucleons in nucleus graph

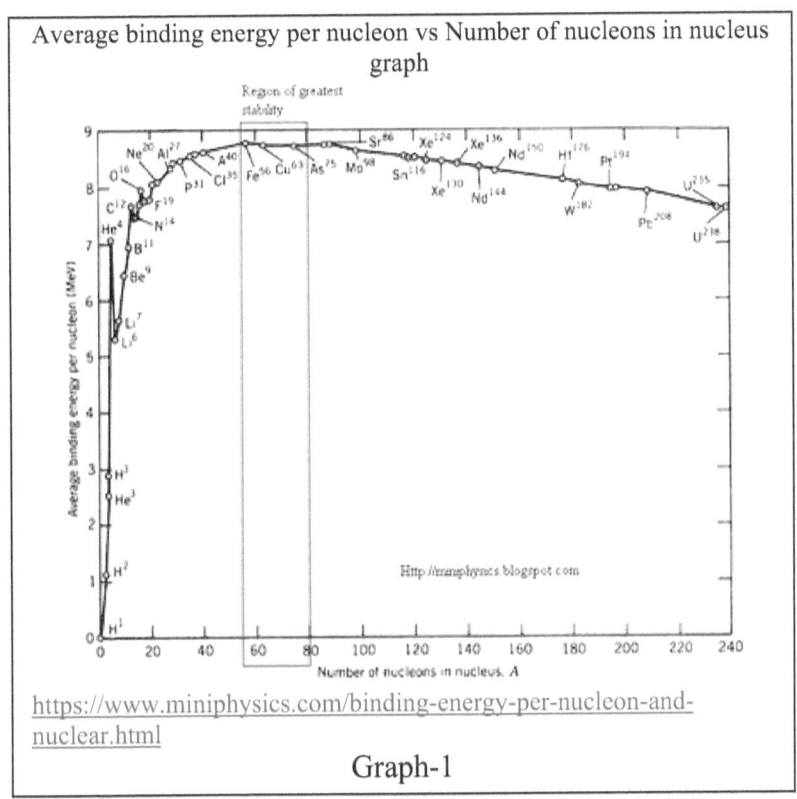

Graph-1

Why nuclear force is attractive?
Nuclear force is by sharing the masses. That is sticking together (attractive). This attractive force spreads around the nucleus.

Why nuclear force is very powerful?
Particles can be attracted to one another in many ways. By sharing part of their masses will be most powerful binding. That is why the **nuclear forces between nucleons is the most powerful forces**.

Why nuclear force is short ranged?
If the attracting particles are spatially apart, the force between them is spread out. In the nucleus, the nucleons are almost fused together in a small space. So, the nuclear force is very much localized in a short distance. That is **the nuclear force is short ranged.**

4. Yukawa's 1935 concept of particle exchange to explain the nuclear force

According to Yukawa, two nucleons exchange particles continuously between then which binds these particles and is the cause of the nuclear force.

Few questions are raised regarding this theory.

1. Why nucleons exchange the mesons? What initiates this exchange?
2. Is this exchange can result in the very high nuclear force?

Yukawa theory creates independent groups of two nucleons in each group and the exchange of particles occurs between the two nucleons of each group. These groups (Yukawa pairs) have no interaction between them and no force between the groups. Thus, a nucleus of multiple nucleons, consists of independent groups of two nucleons each sticking together. There is no force binding all these groups (all nucleons) together. But in reality, a nucleus consists of nucleons all of them sticking together.

Consider an even number n.

If a nucleus has n nucleons, there will be $\frac{n}{2}$ groups of nucleons with two nucleons sticking together in each group. There is no interaction or force between these groups.

In a nucleus of $n + 1$ nucleons, there will be $\frac{n}{2}$ independent groups and 1 nucleon left behind without a partner.

In both these cases, there is no force binding all the nucleons together.

Yukawa put forward his theory of nuclear force and predicted the mass of the exchanging particle (meson). Later a particle of same mass was observed, and Yukawa's theory was taken for granted.

The Yukawa theory of nuclear forces is also called exchange of particle theory. In the exchange of particle theory, a proton emits a π^+ meson and changes to a neutron A neutron absorbs a . π^+ meson and changes to a proton. This kind of changes happens continuously in the nucleus. This kind of transformations changes the atomic number of the particle.

5. Nuclear forces further explained

Nucleons are sticking together by their nuclear defects. Are these shared nuclear defects participating in any activity of the nucleons?

Paper 13 says that the particles are circulating energy packets. Consider 2 nucleons A and B are sticking together by the nuclear force. Shared mass of the nucleon B by the nucleon A (mass defect of A) participates in all activities of the nucleon A except those activities which needs complete independence from other particle. This shared mass participates in the energy circulatory actvity. This further makes the particles A and B bind together.

6. Nuclear force can be calculated based on the explanation given here with the following and related known values.
1. charge of the nucleus and a proton
3. masses of the nucleus. proton and neutron
4. electrostatic force between the nucleus and a proton
5. gravitational force between the nucleus and a neutron
6. nuclear mass defect
7. nuclear force range

PAPER 6.
PHOTON-ELECTRON INTERACTION

Established experimental evidence show that there are 2 types of electrons in a matter – those bound to the matter and those which are free.

Photo electric effect is explained by the absorption of photons by electrons bound to matter. Compton Effect is explained by elastic collision of photons by stationary free electrons in matter.

So matter has 2 kinds of electrons

1. Electrons bound to matter
2. Electrons stationary and free

In photo electric effect, no electron will be ejected from matter below certain incident light frequency (below certain photon energy). This is due to 2 reasons

i. Photon needs at least the minimum energy to overcome the work function needed just to release the electrons bound to the matter
ii. No photon absorption will take place by the stationary free electrons.

Photoelectric equation shows that light consists of particles of energy – photons. The equation does not show any partial absorption of photons. This may be because photons are the smallest packets of energy.

The process of multiple photon absorption by electrons and use these accumulated energies of absorbed photons by electrons to overcome the work function and release electrons is not considered in photoelectric effect. If the absorbed photon energy is less than the work function, it will be released or dissipated by some other way before another photon is absorbed by the same electron.

Further discussion about the work function is given in the paper named "Multiple photon absorption – Photo electric effect".

In Compton Effect, the wavelength of the scattered x-rays is measured. In Compton scattering, two things happenings are –

i. During the scattering of x-rays, the electrons bound to matter will also be emitted like ultraviolet light emit electrons in photoelectric effect.

ii. The shift in the wavelength of scattered x-rays is explained by elastic collision between photons and the stationary free electrons.

From these it follows that –

i The bound electrons will only absorb the photons. There is no elastic collision between bound electrons and photons.

ii There is only elastic collision between stationary free electrons and photons. Stationary free electrons will not absorb photons.

Compton effect is explained by considering light as particles – photons.

PAPER 7.

MULTIPLE PHOTON ABSORPTION – PHOTO ELECTRIC EFFECT

Photoelectric effect was discovered by a by the German physicist Heinrich Rudolf Hertz.in 1887 and a theoretical explanation was given by Albert Einstein in 1905.

On September 27, 1905 when Albert Einstein was employed in the Swiss patent office, Bern, Zurich, working six days a week as a patent clerk published four papers. This paper titled "Does the Inertia of a Body Depend Upon Its Energy-Content?" contained the four papers Photoelectric Effect and Light Quanta, Brownian Motion, Special Relativity and Mass Energy Equivalence.

During Photoelectric effect, photons are absorbed by the electrons bound to matter. That is, an electron completely absorbs one photon. Einstein's photoelectric equation is

$$hf = w + \frac{1}{2} m(v_{max})^2 \quad \text{----} \quad (1)$$

h – Planck's constant
f – frequency of incident light
w – work function of the material on which the light is incident
$\frac{1}{2} m(v_{max})^2$ – maximum kinetic energy of the emitted electrons

This is an equation for one photon absorption. Here an electron absorbs only one photon. So v is replaced by v_1 to show it is due to single photon absorption. So equation (1) is written as

$$hf = w + \frac{1}{2} m(v_{1max})^2 \quad \text{----} \quad (2)$$

By observing the various speeds of electrons emitted from the surface of a material, more information about the binding energies of electrons with that material can be obtained.

All photo electrons are bound to the matter with the same binding energy? Not necessarily. During the photoelectric effect, electrons of different speeds are emitted. Electrons emitted with less kinetic energy (less speed) are those which are bound to the matter with more binding energy. The speed in Einstein's photoelectric equation is the maximum speed of the photoelectrons. So, electrons in Einstein's photoelectric equation are those electrons bound with minimum binding energy (w_{min}) only.

The minimum frequency for the photoelectric effect to occur is the minimum frequency of the incident light for any measurable photoelectrons with the current techniques. Incorporating this, equation (2) is rewritten as

$$hf = w_{min} + \tfrac{1}{2} m(v_{1max})^2 \quad \text{---- (3)}$$

If there are 2 light beams of same frequency are incident from 2 different directions on the surface of the material, an electron bound to the surface can absorb more than one photon and get ejected from the material. If an electron absorbs 2 photons, the photo electric equation should be modified as.

$$2hf = w_{min} + \tfrac{1}{2} m(v_{2max})^2$$

$\tfrac{1}{2} m(v_{2max})^2$ - maximum kinetic energy of the photoelectrons which absorb 2 photons

Multiple photon absorption by an electron is possible

When a photon hit an electron why bound electron absorbs the photon and free electron bounces?

When a photon hits an electron, the immediate response of the electron is to bounce. If the electron is bound to a material, it cannot bounce and the electron absorbs the photon. If the electron is not bound to a material (free electron), the electron is bounced.

PAPER 8.

A WAVE PRODUCED BY AN OSCILLATING OR A VIBRATING OBJECT

The Newtonian (Classical) concept of radiation is that radiation is emitted or absorbed in a continuous wave like form. Max Planck ended classical physics by his quantum theory. In 1900 Max Planck put forward his bold theory that energy is transmitted or absorbed by small packets and not continuously as suggested by the classical physics. The difference between classical and quantum physics is like that of a ramp and a stair case. According to classical physics, events occur in a smooth, orderly, and predictable manner. According to quantum physics, events occur in steps. Explanations of the photoelectric effect (1905) and Compton effect (1923) are the proofs of the quantum theory.

The fact is that, some of the physical phenomena can be explained only by classical wave format, while others can be explained only by quantum theory and some others can be explained both by continuous wave and quantum theory concepts (see table1).

Table 1

PHENOMENON	EXPLAINED BY WAVE THEORY?	EXPLAINED BY QUANTUM THEORY?
Reflection	Yes	Yes
Refraction	Yes	Yes
Interference	Yes	No
Diffraction	Yes	No
Polarization	Yes	No
Photoelectric Effect	No	Yes
Compton Effect	No	Yes

Here is the question. Is energy (ex. light) transmitted in a wave form or as packets of energy (particles)?

In 1924, de Broglie put forward his concept of wave-particle duality. His logic was as follows.
1. Nature loves symmetry
2. Nature exhibits itself in 2 forms – matter and wave.
3. The wave form has 2 aspects – the particle aspect and the wave aspect.
4. Therefore, matter also has 2 aspects - particle aspect and wave aspect.

The wave aspect associated with matter is called matter waves or de Broglie waves.

Now the question is that the classical physics and the quantum physics are really two different ideas?

In this paper, vibration of an object is studied and the resulting disturbance is analyzed. The analysis shows that the disturbance produced by a vibrating object consists of discrete half waves.

These discrete half waves are packets of energy as suggested in the quantum theory. This means classical and quantum theories are not two separate and different theories.

A wave produced by a vibrating object is not continuous.

Consider an object vibrates about its equilibrium point O (Figure 1). It goes to a maximum displacement point A, returns to its original position O, goes to maximum displacement B in the opposite side and comes back to the point O. As the particle moves, every point on its motion produces disturbances which propagates in all directions in the medium with the same speed. The speed of propagation varies from medium to medium.

Consider the particle's movement A to B (A to O to B) and then B to A (B to O to A). This is one vibration (oscillation). These 2 parts of the vibration are represented in Figure1 as follows.

1. Movement A to B (A to O to B) is represented by 1'2'3'.

2. Movement B to A (B to O to A) is represented by 4'5'6'.

Table 2 give the points of disturbances of the particle and the corresponding points on the resulting wave.

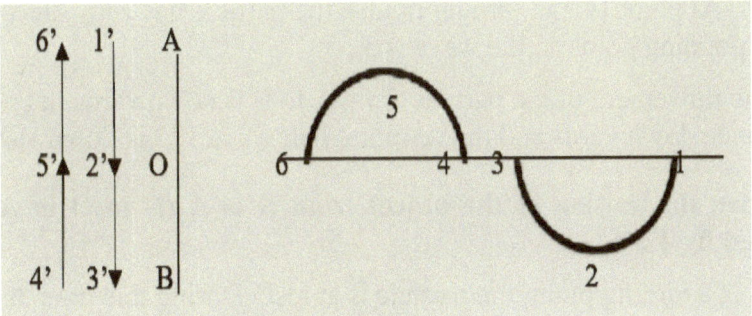

Figure 1.

TABLE 2. POINTS OF DISTURBANCES AND THEIR CORRESPONDING POINTS

	Direction of movement of vibration.					
	A to B			B to A		
Point of disturbance	Corresponding point on its representation 1'2'3'	Corresponding Point on the wave	Point of disturbance	Corresponding point on its representation 4'5'6'	Corresponding Point on the wave
A	1'	1	B	4'	4
O	2'	2	O	5'	5
B	3'	3	A	6'	6

At the maximum displaced points (A and B) where the particle reverses its directions of movement, the speed of the particle will be zero.

1. Consider the motion of the object from A to B (A to O to B) as represented by 1'2'3'.

Since A is a turning point, it starts from rest at A. Corresponding point of the disturbance on the wave is 1. As it moves to point O, the particle picks up speed. The disturbance from the near point of A, moves further than the disturbances from far point from A.

When the speed increases, the disturbance has more energy and corresponding point on the wave has more displacement from the horizontal line.

At point O, the particle has maximum speed (maximum energy) and the corresponding point on the wave has maximum displacement. This is point 2.

When the particle moves to point B, the speed decreases, then the energy decreases and the displacement of the corresponding point on the wave decreases. At point B (the turning point), the particle has zero speed and the corresponding point on the wave is 3.

Since the movement of the particle from A to B is downwards, it creates a disturbance downwards and the resulting half wave 123 is downwards.

2. Consider the motion of the object from B to A (B to O to A) as represented by 4'5'6'.

Since B is a turning point, the particle is at rest. During this time of rest, the half wave 123 has moved further. Then it starts from rest at B and corresponding point of the disturbance on the wave is 4. As it moves to point O, the particle picks up speed. The disturbance from the near point of B, moves further than the disturbances from far point from B.

When the speed increases, the disturbance has more energy and corresponding point on the wave has more displacement from the horizontal line.

At point O, the particle has maximum speed (maximum energy) and the corresponding point on the wave has maximum displacement. This is point 5.

When the particle moves to point A, the speed decreases, then the energy decreases and the displacement of the corresponding point on the wave decreases. At point A (the turning point), the particle has zero speed and the corresponding point on the wave is 6.

Since the movement of the particle from B to A is upwards, it creates a disturbance upwards and the resulting half wave 456 is upwards.

During the motion 1'2'3' a half wave 123 is produced and this disturbance moves out. At the turning point B, the speed is zero and the next half wave 456 corresponding to the motion 4'5'6' is produced a little time after. There is a gap 3-4 between the 2 halfwaves. Thus the 2 half waves are not continuous.

Thus, one vibration of a particle produces 2 separated halves of a full wave.

The Figure2 shows two waves (4 detached half waves) produced by two oscillations of a vibrating particle.

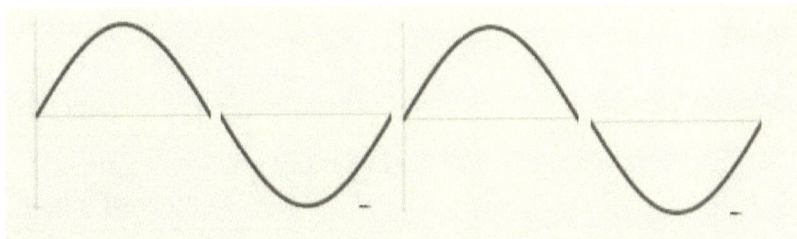

Figure2. Two waves (4 detached half waves) produced by 2 oscillations.

The disturbances move out as separate pulses (detached half waves). Each half wave is an **energy packet.**

Thus, a disturbance produced by a vibrating object travel in the form of a wave which consists of discrete energy packets. **Each half wave is a separate energy packet – half wave energy packet.** The smallest energy packet in nature is a **half wave energy packet.**

The existence of these **half wave energy packets** is explained by the photoelectric and the Compton effects.

These **half wave energy packets** can behave like particles. This particle nature incorporated into the wave nature can explain the physical phenomena given in Table 1.

The relation between these **half wave energy packets** and the energy of a photon is explained in the following discussion of the energy of a photon.

Energy of a Photon

Energy of a photon is $E = hf$

Here h is Planck's constant and f is the frequency of a photon measured in Hertz (Hz). Frequency is the number of waves that passes any point each second. Considering Figure2, frequency is the **number of energy packets** pass any point in one second.

There are two **half wave energy packets** in each wave (see Figure2). So the energy of a photon is the total **half wave energy packets**, 2 times the number of waves passing any point in one second.

That is energy of a photon E = (energy in two **half wave energy packets**) x frequency (number of waves passing any point in one second)

$$= h\ f$$

This is same as in Paper 2

> Measure of a physical quantity = Unit measure x Number of units

If all half waves across the entire energy spectrum have the same energy, the value of h is the same. If the half waves have different energies, the value of h varies.

PAPER 9.
SIGNED PHYSICAL EQUATIONS CONSTANTS

The repulsive force is considered as positive and attractive force is considered as negative. This is the convention. In addition, this sign concept has physical meaning also. In the radial coordinate system, outward direction is the positive r direction, so the repulsive force is positive and attractive force is negative. Further, attractive forces tend to reduce the distance between the objects. Reducing the distance is given a negative sign.

If a sign is given to the constant in the equation of a force between two objects, the resulting force will come with correct sign showing whether it is an attractive or repulsive force.

Force between a negative charge e and a positive charge Q separated by a distance r is given as

$$F = k\frac{(+Q)(-e)}{r^2} \text{ here } \textbf{k is a positive constant}$$

$$= -k\frac{Qe}{r^2}$$

This equation fully describes the force in magnitude and direction. This is a negative force. That is, it is an attractive force.

Force between a 2 negative charges e separated by a distance r is given as

$$F = k\frac{(-e)(-e)}{r^2} \text{ here } \textbf{k is a positive constant}$$

$$= +k\frac{e^2}{r^2}$$

This equation fully describes the force in magnitude and direction. This is a positive force. That is, it is a repulsive force.

Same thing is applicable to the force between magnetic poles also.

If it says that the gravitational force between masses M and m separated by a distance r is,

$$F = G\frac{Mm}{r^2}$$

this equation does not give the direction, the force is not fully described.

If **G is a negative constant**, the force is negative.

$$F = -G\frac{Mm}{r^2}$$

Now the force is fully described in magnitude and direction. This is a negative force. That is, it is an attractive force.

So, the constant of proportionality in the final equation, is a number with a sign.

PAPER 10.
MASS EFFECT AND CHARGE EFFECTS

Consider gravitational and electric forces.

A mass or a charge in space exhibits its presence in its neighborhood. Only another entity which can interact with this effect comes in this neighborhood, the effects of the two entities will interact. The result can be an attractive interaction, a repulsive interaction or no interaction depending on the entities involved.

For example – Consider a mass is placed in a space. This mass exhibits its presence (**mass effect**) in this neighborhood. Only when another mass comes in this space, the two masses interact (attraction). A charge comes in this neighborhood has no interaction with the mass .

Next consider a positive charge is placed in a space. This positive charge exhibits its presence (**positive charge effect**) in this neighborhood. Only when another charge (positive or negative) comes in this neighborhood, there will be interaction between two charges. Interaction of a negative charge will be an attraction and the interaction of a positive charge will be a repulsion. Similar is the case with a negative charge which exhibits its **negative charge effect**.

A fundamental particle will exhibit either its mass effect or its charge effect.

This is similar to the world of living things. An animal produces pheromones and is spread in its neighboring space. It is not attractive or repulsive. It only exhibits its presence. Only animals (of this type) come in this space will be affected by this scent and interaction (attractive or repulsive) takes place.

PAPER 11.
ACCELERATING PARTICLES RADIATE ENERGY

All particles are packets of energies. Some are tightly packed energy, and some are loosely packed energy. The degree of tightness varies from particle to particle.

Energy in a neutral particle is tightly packed and energy in a charged particle is loosely packed.

When a charge is accelerated, it radiates energy (Larmor). This is the result of inertia. When charge is accelerated, it opposes its change of state of motion. The charge will struggle to keep its status quo (state of rest or uniform motion in a straight line). This struggle results in some energy loss from the charge. This is the energy radiated from an accelerated charge.

When a neutral mass is accelerated, opposition of the mass to its change of state of motion also result in a struggle. This should cause the neutral mass to radiate some of its energy.

Particles which are neutral (showing no negative and positive charge) are those with loosely packed equal positive and equal negative charges packed tight which results in zero net charge.

It should be noted that the existence of de Broglie matter waves was experimentally proved by the accelerating charges and the accelerating neutral particles$^+$. Matter waves from the particles is due to the energy dispense of the matter.

Are mass and energy are two entirely different entities? Why accelerating charges and masses radiate energy?

Mass is made up of energy. Mass can be converted to energy. Energy can be converted to mass.

Masses are packets of energy. Energy is not made of particles of masses. Energy under certain conditions forms packet of energy which is mass. So mass is not a separate entity.

When energy in a mass is loosely packed, it is a charged particle. When energy is a mass is tightly packed, the mass is neutral (not electrically charged).

Energy in a mass is linked together. When a mass is accelerated (kicked or jerked) some links may be broken, and some energy is radiated.

If the energy is loosely linked together (in charged particles), more energy is released.

If the energy is tightly linked together (in neutral particles), less energy is released.

[+] https://en.wikipedia.org/wiki/Matter_wave

PAPER 12.
NEWTON'S LAWS OF MOTION

All masses are in any one of the following 3 states of motion.

1. In a state of rest,
2. In motion with constant velocity and
3. In motion with change in velocity (acceleration or deceleration).

All these states of motion are described by Newton's first 2 laws of motion.

Newton's first law of motion is

Scientific breakthroughs that led Newton to his first law of motion are the followings. Galileo Galilei rolled a ball down a slope opposite which was another slope. He found that the ball rolled up the second slope to almost same height which was released on the first slope. In addition, the angle between the slopes has no effect on the height in the second slope. Further he reasonably assumed that if the ball is let go on a plane surface, without the second slope, it will roll indefinitely. It makes sense to see the reason is because the ball never attains the same height as in first slope. Later a French philosopher, mathematician, and scientist (1596-1650), determined that the concept of this continuous motion applies only to motion in a straight line. On this knowledge base, Sir Isaac Newton introduced his first law of motion.

Every object perseveres in its state of rest, or of uniform motion in a right (straight) line, unless it is compelled to change that state by forces impressed thereon. *

* *Principia Mathematica Philosophiae Naturalis*, from W. F. Magie, *A Source Book in Physics* (Cambridge, Mass.: Harvard University Press, 1963).

This first law deals with the first 2 states of motion.

What does this law say? -

i) This **law says** that a mass at rest or in uniform motion in a straight line stays in the same state of motion unless an external force act on it.
ii) This **law also says** that a mass at rest or in uniform motion in a straight line will not change its state of motion by itself. A body at rest or in a uniform motion in a straight line has the ability to stay in this state of motion.

This property of the mass is called inertia.

iii) The law clearly says that the motion of a mass (at rest or in uniform motion in a straight line) will be affected by an external force. The **law does not say** that a mass in any one of the first 2 states of motion will oppose its state of motion when an external force is applied.

This property of mass, which opposes its state of rest or its motion with uniform velocity, for an external force, should be added as another property of the inertia.

Thus, inertia is a property of a mass by which a mass in a state of rest or in a uniform motion with constant velocity will remain in its state of motion and also a mass resists any change of its these two states of motion.

Newton's second law of motion is "Force is equal the change in momentum (mV) per change in time. For a constant mass, force equals mass times acceleration "

If no force is acting, a=0. That is either the mass is at rest or is moving with a uniform velocity. Stating in another words, if no force acts on a mass, the mass will be at rest or keeps its uniform speed in a straight line. **This is Newton's 1st law of motion and is implied in the 2nd law.**

Acceleration will be in the direction of the force. It is verified for linear and circular (nonlinear) motions.

PAPER 13.
SPEED OF LIGHT AND THE ENERGY PER UNIT MASS AND PARTICLES AS PULSATING ENERGY PACKETS

The question is why a photon starts with a speed C and maintain this speed during its journey.

Matter is energy packed together. In some matter, energy is packed tight and, in some others, the energy is packed loose.

When a photon is ejected from a mass, some energy from the mass is converted to a photon and comes out. Energy of a mass m is,

$$E = mC^2$$

In this equation, C is not just the numerical value of the speed of light, but it is the speed of light C associated with the energy.

Equation 2 in Paper2 is

Energy of mass m = Mass (m) x Energy per unit mass (c^2) ---(2)

This energy is kinetic (due to motion). Kinetic energy is related to temperature. The extremely high speed (C) associated with this energy points to a very high temperature. This makes sense since a lot of energy is packed in a small space.

When energy (photon) which was associated with a speed C, is released from a mass, it starts with the same speed C.

This photon is not influenced by any medium and so it moves with the same speed C.

The numerical value of the speed of light is not just some arbitrary number rather it is correlated to some universal fundamentals.

Particles are energy packed together. This energy is kinetic in nature. This kinetic nature of packed energies should explain the positive and negative charges of particles.

Further information about the matter helps for a better understanding of other properties of matter like charge and spin.

Particles as circulating energy packets

Motion is an inherent property of all entities in the universe. The type of universal motion is circulatory in nature like particle spin, electrons motion around the nucleus, planets rotation and motion around sun and so on. All particles are packets of energies. Energy is an active entity. It cannot stay cool, calm, still or inactive. Particles are packet of energies – active and energetic. So, particles should be considered as circulating energy packets.

PAPER 14.

GRAVITATIONAL, ELECTROSTATIC AND NUCLEAR FORCES

Newton's law of gravitational force is derived in this book with the common-sense concept that the effect of a mass at a distance depends on the mass and the distance (see Paper 3).

In a similar manner the equation for electrostatic force between 2 charges is also derived in this book (see Paper 4.1 to 4.3).

These two forces (gravitational and electrostatic) are applicable only for the distances greater than the nuclear range.

The two forces gravitational and electrostatic are of same type. These forces are the forces between 2 particles (masses or charges).

The third force, Nuclear force, is different.

The nucleons are bonded together by sharing (Paper 5.3) masses equivalent to their mass deficiencies. Mass deficiencies are the result of particle shocks (Paper 5.1). This nuclear attractive force, bonding force, is much stronger than the electrostatic repulsive force between two protons separated by a distance.

A nucleon first it has to survive. For example, if a proton comes in nucleus to survive as proton, it should have an exact mass of a proton and other qualities defined by nature. So, it will share a mass with another nucleon to compensate its mass defect. A particle should acquire all the natural characteristics of a proton (like mass, charge etc.) to become a proton. Then only it will exhibit the electrostatic repulsive force between protons. That is, its instinct for existence far supersede its electrostatic properties. The sticking force due to the mass sharing is much higher than the electrostatic repulsion between two protons in the nucleus.

(1) Nucleons contain neutrons and protons.
(2) The distance between particles are within the nuclear range. As a result of the sharing of the mass defect, the nucleons are very close.
(3) The nuclear force is not the force between pairs (two) of nucleons, but it is the force between all the nucleons sticking together (see Paper 5.3).

PAPER 15.

LATENT HEATS OF FUSION AND VAPORIZATION

The heat energy used for phase changes of a substance does not change the temperature of the substance. Since this heat does not increase the temperature, it is called the latent (dormant, inactive) heat. For example, to covert 1 gram of ice at 0°C to 1 gram of water at the same temperature (0°C), an amount of heat of 79.7 calories is required. This is the latent heat of fusion of ice. One gram of water at 100°C needs 533 calories of heat to convert to 1 gram of steam at the same temperature (100°C). This is the latent heat of vaporization of water.

This latent heat is used for what?

Consider the changes during the conversion of

1 gram of ice at -10°C to 1 gram of steam at 100°C.

To see the process at different stages of conversion, divide this into four stages as follows.

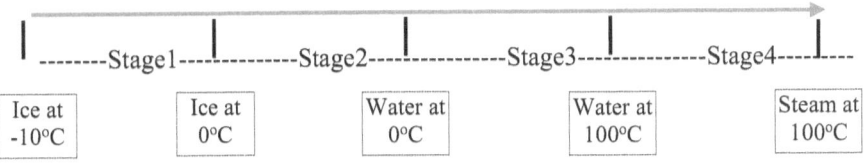

Stage1. Change of ice at -10°C to ice at 0°C.

When heat is given to ice at -10°C, ice warms up to -9°C, then to -8°C, then to -7°C ... then to 0°C. The temperature increases from -10°C to 0°C.

Heat supplied is used for exciting the molecules. The molecules increase their kinetic energies (rotational and/or vibrational). This kinetic energy increase causes the temperature increase. No translation energy is obtained since the molecules are still bound in ice and not moving independently or collectively.

In this stage, heat supplied increases the temperature of the substance (ice).

Stage2. Consider the change of ice at 0°C to water at 0°C.

When heat is supplied to ice at 0°C to change it to water at 0°C, no temperature change occurs.

To start with, the molecules in ice are strongly bound together. Heat is used to loosen the strong attachments of molecules as in ice to lose attachments as in water. This loosening itself will not increase the kinetic energy. When a molecule is loosened, it may gain some kinetic energy. But this gain in kinetic energy is quickly transferred to neighboring molecule to loosen its strong binding to other molecules. This gain in kinetic energy is not recorded as temperature increase since this gain is very localized and lasts only for a very short time. So, the temperature does not change. When whole ice is changed to water at 0°C, binding between water molecules are not as strong as the binding of molecules as in ice. In liquid, they stay together and move and flow freely.

In this stage, heat supplied does not change the temperature of the substance (water).

Heat supplied during this process is the **latent heat of fusion of ice**.

Stage3. Change of water at 0°C to water at 100°C.

Water molecules at 0°C has no strong binding between molecules as in ice. Heat supplied changes the kinetic energy of the water molecules which causes the temperature increase.

In this stage, heat supplied increases the temperature of the substance (water).

Stage4. Change of water at 100°C to steam at 100°C.

Heat supplied is used by molecules to detach completely from one another. Molecules do not get kinetic energy in this process. The detached molecule on the surface escapes to air. If the detached molecules are surrounded by other molecules, they give the excess energy if any to another molecule to detach completely.

In this stage, heat supplied does not change the temperature of the substance (water).

Heat supplied during this process is the **latent heat of vaporization of water**.

PAPER 16.
MEASURED AND EFFECTIVE VALUES IN PHYSICAL EQUATION AND THE CONVERSION FACTOR.

In a physical equation, all the terms should have physical meaning.

In this paper, all letters representing physical values are the measured values. Still, some physical value may be subscripted as measured.

All physical equations can divide into 3 types.

Type1. Equations with proportionality constant is zero.

Type2. Equations with proportionality constant is 1.

Type3. Equations with proportionality constant is other than 0 and 1.

Type1 equation means there is no interaction between the physical quantities

Example of <u>Type2</u> equation is Newton's 2nd law of motion

$$F = ma$$

In this equation, the measured value of mass m and the measured value of acceleration a are applicable. In other words, in Newton's 2nd law of motion,

$$(ma)_{effective} = (ma)_{measured}$$

That is,

$$F = (ma)_{effective} = (ma)_{measured}$$

Example of <u>Type3</u> equation is Newton's law of gravitational force between 2 masses.

$$F = G \frac{m1 \times m2}{d^2}$$

This gives force F between 2 masses $m1$ and $m2$ separated by a distance d.

In this equation, if one applies the product of the measured masses and divide by the square of the measured distance d, the result will not be the F. That is,

$$F \neq \left(\frac{m1 \times m2}{d^2}\right)_{measured}$$

This is because the equation is missing the term G.

The term G usually called as the *constant of proportionality*, is actually a conversion factor (conv. factor).

Conversion factor of a physical quantity in an equation is a constant that converts the measured values to their effective values in the equation.

The effective values of $m1 \times m2$ and/or the effective value of d^2 are not their measured values. Effective and measured values in the law of gravitation are related by the equation,

$$\left(\frac{m1 \times m2}{d^2}\right)_{effective} = G \left(\frac{m1 \times m2}{d^2}\right)_{measured}$$

Where, $m1, m2$ and d are the measured values. In the following equation, a subscript measured is shown to clearly show that these are the measured values $[(\frac{m1 \times m2}{d^2}) = (\frac{m1 \times m2}{d^2})_{measured}]$

$$F = \left(\frac{m1 \times m2}{d^2}\right)_{effective} = conv.factor \times \left(\frac{m1 \times m2}{d^2}\right)_{measured}$$

$$= G \left(\frac{m1 \times m2}{d^2}\right)_{measured} \quad ... \; Eq.\, 1$$

Consider a couple of other examples.

<u>Example1</u>. Electrostatic force between charges $Q1$ and $Q1$ separated by a distance d is

$$F = \left(\frac{Q1 \times Q2}{d^2}\right)_{effective} = conv.factor \times \left(\frac{Q1 \times Q2}{d^2}\right)_{measured}$$

$$= K \left(\frac{Q1 \times Q2}{d^2}\right)_{measured}$$

Con. Factor K is the Coulomb's constant.

<u>Example2</u>. Kinetic Energy of a mass m moving with a velocity v is

$$K.E. = (mv^2)_{effective} = conv.factor \times (mv^2)_{measured}$$

$$= \frac{1}{2}(mv^2)_{measured}$$

Eq. 1 is

$$F = G \left(\frac{m1 \times m2}{d^2}\right)_{measured} \quad \ldots Eq. 1$$

Also

$$F = conv.factor \times \left(\frac{m1 \times m2}{d^2}\right)_{measured}$$

That is, $F = \dfrac{(mass\ conv.factor \times m1)(mass\ conv.factor \times m2)}{(distance\ conv.factor \times d)(distance\ conv.factor \times d)}$

$$= \left(\frac{mass\ conv.factor^2}{distance\ conv.factor^2}\right)\left(\frac{m1 \times m2}{d^2}\right)_{measured} \quad \ldots Eq. 2$$

Comparing Equ.1 and Eq.2, one gets

$$\left(\frac{mass\ conv.factor^2}{distance\ conv.factor^2}\right) = G \quad \ldots Eq.3$$

here G is Universal gravitational constant

In the same way, considering Coulomb's electric force,

$$\left(\frac{charge\ conv.factor^2}{distance\ conv.factor^2}\right) = K \quad \ldots Eq.4$$

here K is Coulomb's law constant.

Here one has 2 equations (Eq.3 and Eq.4) and 3 (mass conversion factor, charge conversion factor and distance conversion factor) unknowns.

All physical equations should have same characteristics. So, the physical equations can be considered as of 2 types. One is **straight terms equations** (with conversion factor equals one) and the second is **converted terms equations** (with the conversion factor other than zero or one).

NOTE:

This conversion factor is applicable in mathematics also.

For example, Area of a circle = **πr^2** here r is the radius of the circle.

In this equation, π is the conversion factor.

PAPER 17.

NEWTON'S SECOND LAW OF MOTION DERIVED FROM NEWTON'S FIRST LAW OF MOTION

Newton's first law is that a body at rest will stay at rest and a body moving with a constant velocity will keep its constant velocity unless the body is acted upon by an unbalanced force.

Newton's second law says $F = ma$, where F is the net force acting on a mass m producing an acceleration a.

There are arguments that Newton's first law can be derived from the second law. Argument is as follows. If the net force F on a mass m is zero, the acceleration a is zero. Zero acceleration can be the body is at rest or moving with a constant velocity – which is Newton's first law.

The first law asserts the existence of inertial frames and the second law does not indicate the existence of inertial frames. An inertial frame of reference is the one in which an object stays at rest or moves with a constant velocity when no net external force acting on it. An inertial frame is a frame that is not accelerating. An accelerating frame of reference is a non-inertial frame. Inertial reference frame is also called inertial frame, Galilean reference frame or inertial space.

Consider some reference frames. First, a reference frame on earth's surface. It is not an inertial frame since earth's rotation about its axis involves acceleration. Second, a reference frame with origin as the center of earth and coordinate axes fixed to distant stars. This is not an inertial frame due to an acceleration due to earth's orbital motion around the sun. Finally, a reference frame with origin as the center of the sun and coordinate axes fixed to distant stars, This is not an inertial frame due to a small acceleration caused by the orbital motion of the sun about the galactic center. *It is almost impossible to identify an absolute inertial frame.*

The intention of this paper is to show that Newton's second law can be derived from his first law.

Newton's second law of motion says that the acceleration a of an object, produced by a net force F, is directly proportional to the force F and inversely proportional to the mass m of the object. Here the constant of proportionality is 1.

That is, $a = \dfrac{F}{m}$

Newton's second law of motion is very elegant. Is this law a basic law or it can be the result (derived from) of another law?

Newton's second law of motion can be explained by Newton's first law of motion.

1. First consider the effect of the mass m on the acceleration and how this effect can be explained by Newton's first law.

   According to the 2nd law, acceleration produced is inversely proportional to the mass of the object.

   $$a \propto \dfrac{1}{m}$$

   This can be explained by the 1st law.

   Let the force F is kept constant.

   Mass is solely dependent on inertia of the object. According to Newton's first law, the inertia is a property of the mass which resists the change in the state of rest or change in the constant velocity motion in a straight line.

   Both these changes involve acceleration. That is inertia is a property of mass which resists acceleration.

   Greater the mass, greater is its inertia and greater is its resisting power. That is, if the mass is increased, acceleration decreases (inversely proportional). If the mass is decreased, the acceleration increases (inversely proportional).

2. Next see the acceleration produced can be explained by Newton's first law. Also, will see how the acceleration produced depends on the force.

   According to the 2nd law, acceleration produced is directly proportional to the force on the object.

   $$a \propto F$$

   This also can be explained by the 1st law.

   Let the mass m is kept constant.

   Newton's first law of motion states that an object at rest stays at rest and an object in motion keeps moving with the same speed in the same direction unless it is acted upon by an unbalanced force.

In the discussion here. consider the second part of the statement. That is, if an object moves with a certain velocity (same speed in same direction), it will keep the same velocity if no net force acting on it.

At any instant let the object moves with a velocity due the force F. Since no other force is acting, the object keeps this constant velocity. Since this steady force F is constantly acting, same velocity is added to the object. This kind of addition of velocity is being applied as long as the force F is in effect. That is the object increases its velocity steadily. This is the acceleration of the object – speeding up. **This is how a steady force causes an object to accelerate. The addition of velocities to the object by a constant force is steady and by the same amount at any equal interval of time. The velocity added increases if the magnitude of the force increases.**

Consider an example – a free falling object under the influence of gravity only, no air resistance or any other force.

Acceleration due to gravitational force is $9.8\ \frac{m}{s^2}$. When the object is dropped from rest, its initial velocity is 0. The constant gravitational force acting on the body gives it some velocity to start. This constant force will keep adding velocities to the object. The equation of motion for the velocity of a free-falling object from rest ($u = 0$) is -

$$v = gt$$

g – acceleration due to gravity and t is time of fall.

This addition of velocities to the object by the constant gravitational force is steady and by the same amount at any equal interval of time. Sum of all these velocities during a period of any 1 sec. of its free fall is $9.8\ \frac{meters}{second}$. During the period of each second, the constant gravitational force adds a speed of $9.8\ \frac{m}{s}$. to the existing speed of the object.

That is during the 1st second of its motion, the object gains a speed of $9.8\ \frac{m}{s}$. During the 2nd second of its motion, the object gains an additional speed of $9.8\ \frac{m}{s}$. and so on. This is shown in a table here.

VELOCITIES OF A FREE-FALLING OBJECT FROM REST

AT THE END OF	VELOCITY $(\frac{meters}{second})$
At start	0
$\frac{1}{100}$ th of a second	$\frac{9.8}{100}$
$\frac{1}{10}$ th of a second	$\frac{9.8}{10}$
$\frac{1}{n}$ th of a second	$\frac{9.8}{n}$
1ST sec.	9.8 x 1
2nd sec.	9.8 x 2
3rd sec.	9.8 x 3
4th sec.	9.8 x 4

PAPER 18.

GENERAL EQUATIONS OF MOTION AND THE EQUATION OF THE CONSERVATION OF MECHANICAL ENERGY IN A FREE FALL.

Free fall of a mass under gravity without any other force is considered in this paper. Two sets of equations applicable in a free fall, the general equations of motion and the conservation of total mechanical energy equation, are discussed. Conservation of mechanical energy is further analyzed.

Next, the relationship between these two sets of equations is studied. It is shown that these two sets are closely related. Further it is demonstrated that the general equations of motion can be derived from the conservation of mechanical energy equation and vice versa.

The two sets of equations which are applicable to a free-falling object are the following.

1. THE GENERAL EQUATIONS OF MOTION (TABLE-1)

Table-1

NO.	GENERAL EQUATIONS OF MOTION		
	GENERAL	UNDER GRAVITY	
		UNDER GRAVITY (a=g)	UNDER GRAVITY STARTS FROM REST (u=0)
1	$v = u + at$	$v = u + gt$	$v = gt$
2	$s = ut + \frac{1}{2}at^2$	$s = ut + \frac{1}{2}gt^2$	$s = \frac{1}{2}gt^2$
3	$v^2 = u^2 + 2as$	$v^2 = u^2 + 2gs$	$v^2 = 2gs$

u is the initial velocity, v is the velocity at any time t, a is the acceleration, s is the distance travelled (displacement) in time t and g is the acceleration due to gravity.

2. CONSERVATION OF TOTAL MECHANICAL ENERGY EQUATION

If a body of mass *m* falls from a height above the ground without any outside force like air resistance, the total mechanical energy at all points on its path will be the same. For a mass m at a height h above ground with a velocity v,

Total Mechanical Energy = Gravitational Potential Energy + Kinetic energy

$$= mgh + \frac{1}{2}mv^2$$

Energy is expressed in Joules.

CONSIDER A DIMENSIONAL ANALYSIS OF GRAVITATIONAL POTENTIAL ENERGY AND THE KINETIC ENERGY.

Gravitational Potential Energy $mgh \rightarrow m \frac{v}{t} l$

$$m \frac{l}{t} \frac{1}{t} l \rightarrow m \frac{l^2}{t^2} \quad \dots\dots\dots\dots\dots (A)$$

Kinetic Energy $\frac{1}{2}mv^2 \rightarrow \frac{1}{2}m\frac{l^2}{t^2}$ ...... (B)

Dimensions of A and B are the same.

Here a factor ½ appears in equation B.

Question is what is this ½ doing on the right side of the kinetic energy equation?

This factor ½ is the conversion factor (see Paper 16). That is, all the terms in a physical equation has its physical meaning and should be explained. The measured values of the mass and the velocity in the KE equation are not the effective values coming in the equation. This conversion factor $\frac{1}{2}$ changes these measured values to its effective values.

CONVERSION FROM POTENTIAL ENERGY (PE) TO KINETIC ENERGY (KE).

When a body falls down in a free fall, the gravitational potential energy is reduced (as the height decreases) and the kinetic energy increases by the same amount. **This can be thought of due to two nonrelated events.**

1. As the potential energy decreases, in order to conserve the total mechanical energy, the reduction in PE is converted to KE. This increases the velocity appropriately, so the kinetic energy increases by the same amount.

   If a mass is held at a height from the ground, the energy (gravitational potential energy) is stored in the holding device. No energy is stored in the mass. **The mass has no device or method to store this energy.** When the object is released, the energy is transferred from the holding device to the mass.

2. For an object falling freely, due to the constant gravitational acceleration, the velocity increases.

What happens is the following.

As the mass is accelerated downward due to gravity, its velocity increases. This does work on the mass.

$$Work\ done = W = F\ \Delta s = m\ a\ \Delta s$$

Here a is the acceleration and Δs is the distance travelled.

This work done is the kinetic energy and is stored in the mass. Using the simple algebra, this kinetic energy can be calculated as $\frac{1}{2}mv^2$

That is as the mass falls, force of gravitation does work on the mass, and the work done. This work done is equivalent to the reduction in its potential energy and is changed to kinetic energy. That is the reduction in potential energy is transferred to the mass as kinetic energy.

In the potential energy (mgh), g is the gravitational pull. In the kinetic energy ($\frac{1}{2}mv^2$), v depends on the gravitational acceleration. Thus, in a free fall, g is the link between these 2 types energies

Question is whether the two sets of equations (conservation of total mechanical energy equation and the general equations of motion) are independent or related for the motion of an object under gravity? Their relationship can be investigated in different ways.

One method is to obtain the numerical values of physical quantities using both sets of equations and see if they are the same.

Another method is to calculate the heights and velocities for different times from the general equations of motion. Then apply these values to get the total mechanical energy and see whether it is conserved. Three cases of this method are shown in the following tables.

It is found the total mechanical energies calculated using h and v values obtained from the general equations of motion at different times during the motion of an object are the same. That is, the total mechanical energy is conserved. Following 3 cases (each case in separate table – Table-2, Table-3 and Table-4) show these calculations. The last column in each table is the total mechanical energy for that system and it is constant for different times which are calculated from the height and velocity values obtained from the equations of motion*. Positive values of the PE are entered in the tables.

Table-2 (Case 1)
A mass of 1 kg is dropped from a height of 70 meters with an initial velocity of 15 m/s
(All quantities to up direction are positive and down direction are negative.
So, g = -9.8 m/s² and u = -15 m/s)

Time (sec.)	Use general equations of motion			Calculation of total mechanical Energy ME = PE + KE (all are in Joules)		
	Distance travelled $s = ut + \frac{1}{2}gt^2$ (meters)	Velocity $v = u + gt$ (meters/ second)	Height (distance from ground) $h = [70 - |s|]$ (meters)	Potential Energy (gravitational) $|PE| = mgh$	Kinetic Energy $KE = \frac{1}{2}mv^2$	Total Mechanical Energy ME=PE+KE
0	0	-15	70	686	112.5	798.5
1	-19.9	-24.8	50.1	490.98	307.52	798.5
1.5	-33.5	-29.7	36.47	357.41	441.05	798.5
2	-49.6	-34.60	20.4	199.92	598.58	798.5

Table-3 (Case 2)

A mass of 1 kg is dropped from rest from a height of 70 meters
(All quantities to up direction are positive and down direction are negative.
So, g = -9.8 m/s² and u = 0)

	Use modified general equations of motion for initial velocity u = 0			Calculation of total mechanical Energy ME = PE + KE (all are in Joules)		
Time (sec.)	Distance travelled $s = \frac{1}{2}gt^2$ (meters)	Velocity $v = gt$ (meters / second)	Height (distance from ground) $h = [70 - |s|]$ (meters)	Potential Energy (gravitational) $|PE| = mgh$	Kinetic Energy $KE = \frac{1}{2}mv^2$	Total Mechanical Energy ME=PE+KE
0	0	0	70	686	0	686
1	-4.9	-9.8	65.4	637.98	48.02	686
1.5	-11.03	-14.7	58.97	577.91	108.05	686
2	-19.6	-19.6	50.4	493.92	192.08	686
3	-44.1	-29.4	25.9	253.82	432.18	686

Table-4 (Case 3)

A mass of 1 kg is projected straight up with an initial velocity of 50 m/s
(All quantities to up direction are positive and down direction are negative.
So, g = -9.8 m/s² and u = 50 m/s)

	Use general equations of motion			Calculation of total mechanical Energy ME = PE + KE (all are in Joules)		
Time (sec.)	Distance travelled $s = ut + \frac{1}{2}gt^2$ (meters)	Velocity $v = u + gt$ (meters / second)	Height (distance from ground) $h = s$ (meters)	Potential Energy (gravitational) $|PE| = mgh$	Kinetic Energy $KE = \frac{1}{2}mv^2$	Total Mechanical Energy ME=PE+KE
0	0	50	0	0	1250	1250
1	45.1	40.2	45.1	441.98	808.02	1250
2	80.4	30.4	80.4	787.92	462.08	1250
3	105.9	20.6	105.9	1037.82	212.18	1250
4	121.6	10.8	121.6	1191.68	58.32	1250

* *Calculations are performed by Jensen Jacob*

From above discussions, it seems that the two sets of equations are not independent. Here, conservation of mechanical energy equation is derived from equations of motion and two equations of motion are derived from the conservation of mechanical energy equation.

There are 2 parts following this.

Part 1 is the derivation of conservation of mechanical energy equation from all three general equations of motion given in Table 1.

Part 2 is the derivation of two of the general equations of motion given in Table 1 from conservation of mechanical energy equation.

Part 1. Derive the equation of conservation of mechanical energy from general equations of motion

1. Consider the first general equation of motion in Table-1.

$$v = u + at$$

Consider an object falls from rest ($u = 0$) from a height H and v is its velocity at the bottom (Figure-1).

Since u=0 and a=g (acceleration due to gravity),

$$v = gt$$
$$v^2 = g^2 t^2$$

Multiplying both sides by $\frac{1}{2} m$,

$$\frac{1}{2} m v^2 = \frac{1}{2} m g^2 t^2$$

Substituting for one value of g ($g = \frac{v}{t}$),

$$= \frac{1}{2} m g \frac{v}{t} t^2$$

That is, $\quad \frac{1}{2} m v^2 = m g \frac{1}{2} v t$ eq.1

Since $u = 0$, $\frac{1}{2} v t = \frac{(v+u)}{2} t = v_{average} \, t = H$.

So, eq.1 becomes,

$$\frac{1}{2} m v^2 = m g H$$

$$m g H = \frac{1}{2} m v^2$$

This is the equation of conservation of mechanical energy.

When a body starts from rest from a height H, and arrives at the bottom with a velocity v,

Total mechanical energy on top (potential energy) = Total mechanical energy at the bottom (kinetic energy)

Figure-1

2. Consider the second general equation of motion in Table-1.

$$s = ut + \frac{1}{2}at^2$$

Consider an object falls from rest (u=0) from a height H and v is the velocity at the bottom (Figure-2).

$$\text{since } u = 0, \quad s = \frac{1}{2}at^2$$

$a = g$ (acceleration due to gravity). When the object reaches the ground, the distance travelled $s = H$.

$$\text{so, } H = \frac{1}{2}gt^2$$

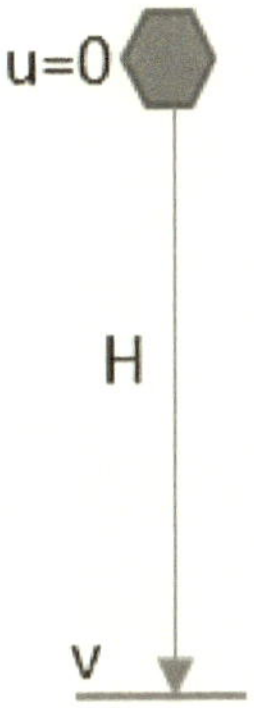

multiplying both sides by mg,

$$mgH = \frac{1}{2}mg^2t^2$$

If v is the velocity at the bottom and t is the time taken to reach the bottom, $g = \frac{v}{t}$.

$$mgH = \frac{1}{2}m\frac{v^2}{t^2}t^2$$

$$mgH = \frac{1}{2}mv^2$$

This is the conservation of mechanical energy equation.

When a body starts from rest from a height H, and arrives at the bottom with a velocity v,

Total mechanical energy on top (potential energy) = Total mechanical energy at the bottom (kinetic energy)

Figure-2

3. Consider the third general equation of motion in Table-1.

$$v^2 = u^2 + 2as$$

Consider an object falls from a height H with an initial velocity u (Figure- 3).

At a distance s from top, let the velocity of the object be v. An equation of motion for this object is,

$$v^2 - u^2 = 2as$$

Since $a = g$ (acceleration due to gravity),

$$v^2 - u^2 = 2gs$$

Since distance travelled s = H-h (see Figure-3),

$$v^2 - u^2 = 2g(H - h)$$
$$v^2 - u^2 = 2(gH - gh)$$

Dividing both sides by 2,

$$\frac{1}{2}v^2 - \frac{1}{2}u^2 = gH - gh$$

Multiply both sides by m,

$$\frac{1}{2}mv^2 - \frac{1}{2}mu^2 = mgH - mgh$$

rearranging, $mgH + \frac{1}{2}mu^2 = mgh + \frac{1}{2}mv^2$

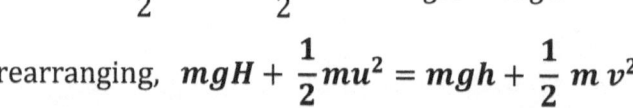

Figure-3

This is the conservation of mechanical energy equation.

Total mechanical energy at start =
                      Total mechanical energy at any other point.

Part 2. Derive general equations of motion from the equation of conservation of mechanical energy.

Applying a reverse procedure for 1 and 3 in Part 1, 1 and 2 of Part 2 are obtained.

1. Derive a modified first general equation of motion in Table-1 from the conservation of mechanical energy equation.

Consider a free-falling object from rest ($u = 0$) of mass m from a height H. Let v is the velocity of the object as it reaches the ground (Figure- 4).

According to the conservation of mechanical energy,

Total mechanical energy at height H = Total mechanical energy at the bottom.

That is $mgH = \frac{1}{2} m v^2$

g is the acceleration a, H is the distance travelled s and cancel m from both sides

$$as = \frac{1}{2} v^2$$

$$s = v_{average}\, t = \frac{(v + 0)}{2} t = \frac{v}{2} t, \text{ since } u = 0$$

$$\text{So,} \quad a\, \frac{v}{2} t = \frac{1}{2} v^2$$

$$a t = v$$

That is, $v = at$

This is the modified (for u=0), first general equation of motion in Table- 1 derived from the conservation of mechanical equation.

2. Derive the third general equation of motion in Table-1 from the conservation of mechanical energy equation.

In a free fall of an object of mass m, consider two heights h_1 and h_2. Let u is the initial velocity (at height h_1) and v is the velocity after it travelled a distance s (at height h_2) see Figure-5. According to the conservation of mechanical energy, total mechanical energy at these heights will be the same.

$$mgh_1 + \frac{1}{2}mu^2 = mgh_2 + \frac{1}{2}mv^2$$

Rearranging, $mg(h_1 - h_2) = \frac{1}{2}m(v^2 - u^2)$

Dividing both sides by m,

$$g(h_1 - h_2) = \frac{1}{2}(v^2 - u^2)$$

since $(h_1 - h_2) = s$, $\quad gs = \frac{1}{2}(v^2 - u^2)$

$$2gs = (v^2 - u^2)$$

$g = a$ (acceleration).

So, $v^2 - u^2 = 2as$

That is, $v^2 = u^2 + 2as$

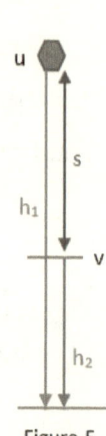

Figure-5

This is one way of deriving the third general equation of motion from the conservation of mechanical energy equation.

Conversions between these two sets of equations (general equations of motions and conservation of mechanical energy equation) discussed in this paper can be done different ways.

For example, consider an object is projected up with an initial velocity u. Let its velocity at a height h is v as shown in Figure-6. Now the third general equation of motion in Table-1 can be derived from the conservation of mechanical energy equation.

According to the conservation of mechanical energy, total mechanical energy at start and at height h will be the same.

That is, $\frac{1}{2} m u^2 = mgh + \frac{1}{2} m v^2$

$$\frac{1}{2} u^2 = gh + \frac{1}{2} v^2$$

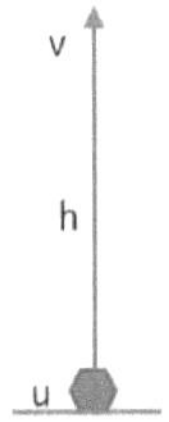

Here g is the acceleration a and h is the distance s travelled,

$$u^2 = 2as + v^2$$

Figure-6

Since the acceleration a (due to gravity) is opposite to the directions of the velocities,

$$u^2 = -2as + v^2$$

Rearranging, $v^2 = u^2 + 2as$

This is the third general equation of motion in Table-1 derived previously from a different situation (see 2 in Part 2).

PAPER 19.

GENERAL EQUATIONS OF MOTION AND THE EQUATION OF THE CONSERVATION OF MECHANICAL ENERGY IN A PENDULUM MOTION.

The motion of a simple pendulum motion is considered in this paper. A simple pendulum is a mass attached to a non-stretchable string. The other end of the string is attached to a fixed point and the pendulum is allowed to move back and forth in a vertical plane. The ideal situation with no resistance to the motion is considered here. At its maximum displacement position A, let h be its height from the horizontal and B is the equilibrium position of the pendulum mass as shown in Figure-1.

Conservation of total mechanical energy equation is applicable in a simple pendulum motion. In this paper. it is investigated whether the general equations of motion is applicable in the motion of a simple pendulum.

General equations of motion are given in Table-1.

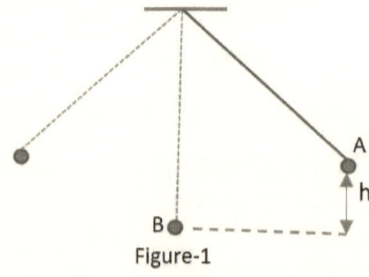

Figure-1

Table-1

NO.	GENERAL EQUATIONS OF MOTION		
	GENERAL	UNDER GRAVITY	
		UNDER GRAVITY (a=g)	UNDER GRAVITY STARTS FROM REST (u=0)
1	$v = u + at$	$v = u + gt$	$v = gt$
2	$s = ut + \frac{1}{2}at^2$	$s = ut + \frac{1}{2}gt^2$	$s = \frac{1}{2}gt^2$
3	$v^2 = u^2 + 2as$	$v^2 = u^2 + 2gs$	$v^2 = 2gs$

u is the initial velocity, v is the velocity at any time t, a is the acceleration, s is the distance travelled (displacement) in time t and g is the acceleration due to gravity.

Consider the motion of the pendulum mass from position A (a maximum displacement position) in the absence of any resistance to its motion. When the mass comes to its mean (equilibrium position) position, let its velocity be v.

According to the conservation of mechanical energy,

Total mechanical energy at A = total mechanical energy at B.

$$\text{That is,} \quad mgh = \frac{1}{2}mv^2$$

$$\text{So,} \quad v = \sqrt{2gh} \quad \text{............eq.1}$$

The 3rd general equation of motion in Table-1 is,

$$v^2 = u^2 + 2as$$

Since $u = 0$ (at maximum displacement position) and $a = g$ (acceleration due to gravity),

$$v = \sqrt{2gs} \quad \text{............eq.2}$$

From eq. 1 and eq. 2, the distance travelled s in the 3rd general equation of motion should be the vertical height.

Applying the 1st general equation of motion in Table-1,

$$v = u + at$$

Since $u = 0$ and $a = g$ (acceleration due to gravity),

$$v = gt$$

Substituting for v from eq.1,

$$\sqrt{2gh} = gt$$
$$2gh = g^2 t^2$$
$$2h = gt^2$$
$$t = \sqrt{\frac{2h}{g}} \quad \text{................... eq.3}$$

The 2nd general equation of motion in Table-1 is,

$$s = ut + \frac{1}{2}at^2$$

Again since $u = 0$ and $a = g$ (acceleration due to gravity),

$$s = \frac{1}{2}gt^2$$

Substituting for t from eq.3,

$$s = \frac{1}{2}g\frac{2h}{g}$$

That is, $s = h$ eq.4

These discussions show that the general equations of motion is applicable for a simple pendulum motion if the distance s is the vertical height of the pendulum mass from the horizontal passing through the equilibrium position of the pendulum mass.

PAPER 20.

MEASURE THE SPEED OF EARTH AROUND SUN

A frame of reference is a set of coordinates that can be used to determine positions and velocities of objects in that frame. A reference frame is also called reference system or a coordinate system. The two types of reference frames are inertial and non-inertial.

An inertial reference frame is one in which Newton's first law of motion, which defines inertia (a property of matter) is valid. An inertial frame is a reference frame at rest or moving with a constant velocity in a straight line. A non-inertial frame is one which is accelerating. All measurements or observations will be compared to a reference frame. Most used reference frame is the Earth. Earth is not strictly an inertial reference frame because of its nonlinear motion around sun. If one considers only a small time of Earth's motion around sun, earth is moving almost in a straight line and is almost an inertial frame.

The Earth is moving around the sun with a high speed (about 67,000 miles per hour - 30 kilometers per second). An observer on earth has no sensation of this motion. Furthermore, it is assumed that an experiment conducted in an inertial frame cannot detect the motion of that inertial frame. This topic will be discussed in this paper.

Consider a train ABCD moving in a straight line with a constant velocity u, as shown in Figure-1. This will be an inertial reference frame. In this inertial frame, the path of a free-falling object is studied. An object is dropped from a point X at a certain height from the floor. For an observer in the train, the path of the object is vertically down XY. Since the train is moving while the object falls, the train is moved to $A_mB_mC_mD_m$, the point X is moved to X_m and the point Y is moved to Y_m. While this body is falling, the train is moving. So, for an observer outside the train, the object takes a different path XY_m and falls at point Y_m, (moved location of the point the Y) as shown.

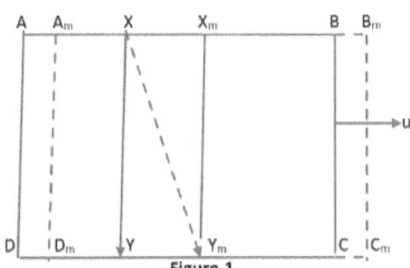

Figure-1

For an observer in the train, the object appears to be falling straight down perpendicular to the ground and it does not take a slanted path.

For an observer outside the frame the motion of the object will not be falling straight down. If u is the velocity of the train and t is the time of fall of the object, the actual path travelled XY_m by the object can be shown as in Figure-1a. If the train is moving to the left instead of moving to right as in Figure-1, the actual path travelled XY_m by the object can be shown as in Figure-1b. Here the equation $s = ut + \frac{1}{2}gt^2$ is used.

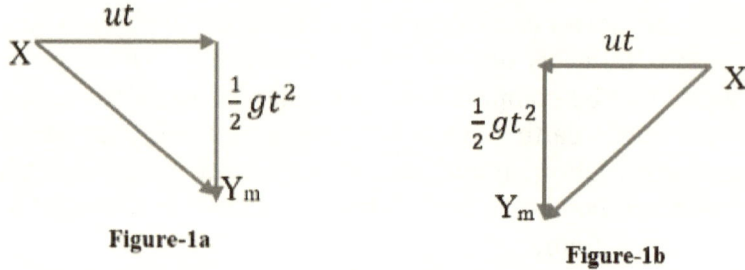

Figure-1a   Figure-1b

The argument of why the object appears to be falling straight down to an observer in the train is the following. The object is moving with the same velocity as the frame immediately before it is dropped. Then, it will continue to move in the same direction with the same velocity unless an external force act on it (Newton's 1st law of motion).

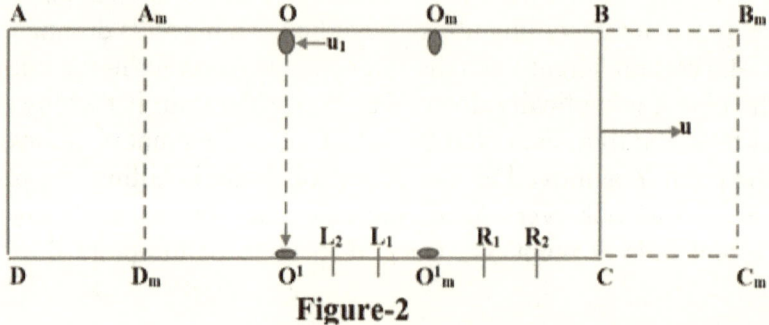

Figure-2

Considering the above explanation of a free-falling object in a train moving along a straight line with a constant velocity, question is can one detect the train motion by an experiment conducted in the train?

In Figure-2, ABCD is a train moving to right in a straight line with a constant velocity u as shown.

To start with mark a point X vertically bellow the object O on the train floor. All the impact points in this experiment should be measured relative to this point X.

An object O is dropped from a height in the train. The point O^1 is on the train floor and is vertically below O. By the time the object reaches the train floor, the train is moved to $A_m B_m C_m D_m$, point O is moved to O_m and the point O^1 is moved to O^1_m. When the object makes a free-fall, the observer in the train sees the object hits at O^1_m (the point X).

Some facts about free falling objects relevant to this paper are –

1. The path of the free-falling object is greater if the object has a horizontal velocity than the path of a free-falling object with no horizontal velocity.

2. The time taken by two objects (with and without a horizontal velocity) from the same height to reach the ground will be the same. This is because the horizontal component of the velocity does not affect the vertical component of the velocity. If there is no vertical component of a velocity, the time t an object takes to reach the bottom from a height h is

$$t = \sqrt{\frac{2h}{g}}$$

Here g is the acceleration due to gravity.

So, in Figure-2, if the object O falls freely is given an initial horizontal velocity to the opposite direction of the train motion, the net horizontal velocity of the object in the train direction is less than the train velocity and it will hit on the left side of the mark X. And if this object is given an initial horizontal velocity in the direction of the train motion, the net horizontal velocity of the object in the train direction is more than the train velocity and, it will hit to the right side of the mark X.

This is illustrated in Figure-1ca, Figure-1cb and Figure1cc. In these figures, u is the velocity of the frame, AB represents the net initial horizontal velocity of the falling object in magnitude and direction (roughly) and t is the time of fall.

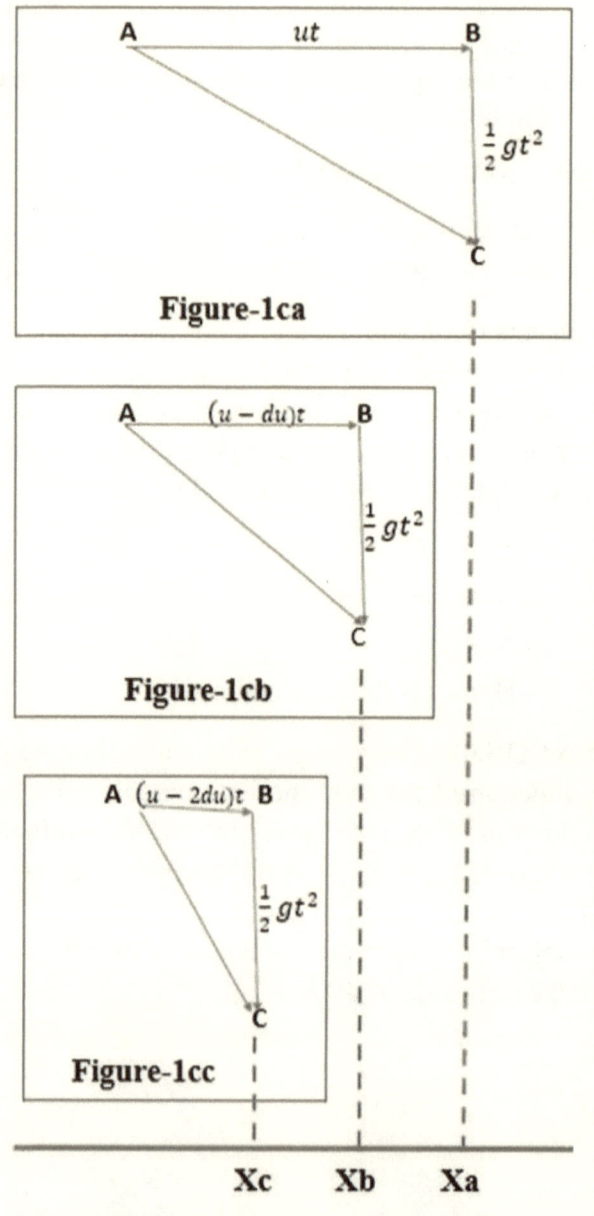

In Figure-1ca, the falling object is not given any additional velocity. Its initial horizontal velocity is the velocity u of the frame So AB=ut.

In Figure-1cb, the falling object is given an initial velocity du in the opposite direction of the train velocity. So, AB= $(u-du)t$

In Figure-1cc, the falling object is given an initial velocity $2du$ in the opposite direction of the train velocity. So, AB= $(u-2du)t$

As one can see from these figures, the impact point of the object moves to left as its initial horizontal velocity in the direction of the train motion decreases –

for the velocity u the object falls at X_a, for the velocity $(u-du)$ the object falls at X_b and for the velocity $(u-2du)$ the object falls at X_c.

4. The distance between the points X and the point where the object hits the floor is related to the net horizontal velocity and the time of flight of the object.

Consider the following 3 cases.

Case-1. Give the falling object a small velocity, less than the velocity V of the train, in a direction opposite to the train motion.

Give the object a velocity v_1 (directed to the left) opposite to the train motion as shown. Start with a value for u_1 little less than u. For example, if u=20 m/s, u_1=18m/s. Then let the object fall. The train moves with velocity u to right and the object falls with a lesser velocity (u – u_1) in the train's direction. According to numerical example given, the train velocity in this case is u=20m/s and the object velocity is (u-u_1) = (20m/s-18m/s) = 2m/s. The observer in the train will see the object falls on the left side of O'_m (left side of the mark X) say at L_2. Now reduce u_1 (say to 16m/s). By the same argument the object will hit on the left side of O'_m say L_1. That is as long as the object's velocity is less than u, the result is the object is hitting the floor to the left side of O'_m. The location of O'_m is marked as X before.

Case-2. The falling object has a velocity in the train direction same as the velocity of the train.

This is the normal situation. When u_1=0, the velocity of the object in the train direction will be the same as the velocity of the train. Now the observer in the train sees the object falls vertically down to O'_m (at point X).

Case-3. Give the falling object, a small velocity, less than the velocity u of the train, in the same direction as the train motion.

Reverse the direction of u_1 to the right (to the direction of the train motion). Start with a small value for u_1. Now object has a velocity (u+u_1) in the train direction. In this case, the object is moving in the direction of the train and has a velocity greater than the train velocity. The observer will see the object falls to the right side of O'_m say R_1. Remember the location of O'_m is marked as X to start with. As u_1 is increased, the object seems to hit R_2 and so on.

This change of impact point from left to right of O'_m, (the point marked as X) to an observer in the train shows the train is moving with a velocity u to right. This clearly proves the constant velocity of the train along a straight line.

Thus, motion of an inertial frame (a frame moving with a constant velocity in a straight line) can be detected and its velocity can be measured.

Earth is almost an inertial frame when consider its motion around sun for a short period of time. The experiment given here can be modified and used to detect and measure Earth's speed around sun.

PAPER 21.
A STATIONARY PENDULUM IS A SUBSTITUTE FOR FOUCAULT PENDULUM

A Foucault pendulum is a simple pendulum specially designed to swing back and forth in any direction. It is unlike a grandfather clock which has a pendulum restricted to swing only in a specific plane. The pivot is special in that it must allow the string to swing in any direction with exactly equal ease, and without transmitting the rotation of the attachment point to the pendulum. The plane in which Foucault's pendulum swings takes a plane in space not related to Earth's reference system (a coordinate system fixed on earth). So as the Earth rotates, the pendulum plane does not rotate. For an observer on earth, it will look like the pendulum plane rotates as earth rotates about its axis.

Consider an observer is on the North Pole of the Earth and look at the swing of the Foucault pendulum. In about 24 hours, The Earth rotates once (rotates 360 degrees) about its axis. Now, since the plane in which pendulum swing does not change with the rotation of the earth, to this observer, the pendulum swing plane looks rotated one full circle (360 degrees). If the observer goes to lower latitudes (away from North Pole), the pendulum seems to take more than 24 hours to complete one full circle. At the equator, the pendulum has no apparent rotation at all.

The only reason for a swinging pendulum is to define a plane fixed in space independent of the earth's reference frame. Physical constants like length, mass or angle of rotation have no relation with the apparent rotation of the pendulum swing plane.

Therefore any stationary pendulum (any stationary hanging plane) designed to rotate freely in any direction can substitute for a Foucault Pendulum. If one can set up a plane in space independent of the Earth's coordinate system, one can see the apparent rotation of the plane due to earth rotation.

The author designed a crude set up as shown in Figure-1 to study the rotation of a frame and its independence on a stationary pendulum suspended in this rotating frame.

ND-P200T motorized turntable.
Circular wood placed on the turn table.
An L shaped metal suspended from a frame attached to the circular Wood.

Figure-1

A ND-P200T motorized turntable which has about 1.5 rotations per minute was used in the experiment. When the turntable turns, the observer sees the circular wooden platform resting over the turn table and the attachments on the wooden platform turns. A stationary L shaped metal suspended has a plane independent of the rotating table reference frame. It is observed that the plane of the suspended object is nearly stationary and does not turn with the rotating circular wooden platform. There is a lot of inaccuracies in this experiment like very rude pivot and other conditions.

PAPER 22.

GENERAL EQUATIONS OF MOTION ARE THE SUM OF THE PARTS OF INITIAL VELOCITY AND THE ACCELERATION

A motion is a state of change of position of an object in space over time. The motion is described in terms of displacement, velocity, acceleration and time. The relationship between these quantities is called equations of motion.

The three general equations of motion for uniform (constant) acceleration are -

$$v = u + at \quad \text{.................. A}$$
$$v^2 = u^2 + 2as \quad \text{........... B}$$
$$s = ut + \frac{1}{2} a t^2 \quad \text{......... C}$$

Here, s is the displacement, u is the initial velocity, v is the final velocity, a is the acceleration and t is the time of motion. These equations are applicable only for an object with constant acceleration in a straight line.

These equations can be derived using simple algebra. These equations are also referred as SUVAT equations. Each letter in this represents the physical quantities involved in the general equations of motion.

Procedure used to get these equations of motion by algebra is the following.

$$acceleration = \frac{final\ velocity - initial\ velocity}{time}$$

That is, $a = \frac{v-u}{t}$

$at = v - u$

So, $v = u + at$ This is equation A

Now, from equation A, $t = \frac{(v-u)}{a}$ 1

$distance = average\ velocity \times time$

$$s = \frac{final\ velocity + initial\ velocity}{2} \times t$$

That is, $s = \frac{v+u}{2} \times t$

Substituting for t from eq. 1, $s = \dfrac{v+u}{2} \times \dfrac{v-u}{a}$

$$2as = v^2 - u^2$$

$$v^2 = u^2 + 2as \quad \text{This is equation B}$$

Distance travelled $s = $ Average velocity $\times$ time

$$= \dfrac{(v+u)}{2} \times t$$

Substituting for v from equation A,

$$s = \dfrac{(u + at + u)}{2} \times t$$

That is, $s = \dfrac{(2u + at) \times t}{2}$

$$s = ut + \dfrac{1}{2} at^2 \quad \text{This is equation C}$$

In this paper these equations are shown to be the sum of two parts – initial velocity and the acceleration. The first term of these equations is the component of initial velocity and the second term is the component of the constant acceleration. So, the general equations of motion are the sum of

1. Contribution from the initial velocity
   and
2. Contribution from the acceleration

1. The first general equation of motion is $v = u + at$ .... A

The contribution to the final velocity v due to the initial velocity is u which does not change.

So, Contribution to the final velocity v due to initial velocity $= u$ A1

Contribution to the final velocity v due to acceleration is,

$$at = \dfrac{v_a}{t} t$$

$$= v_a \quad .... A2$$

Here v_a is the velocity contribution due to the acceleration.

Substituting for at from A2, equation A becomes,

$$v = u + v_a \quad .... D$$

That is,

Final velocity of the object = contribution from the initial velocity + contribution from the acceleration.

2. The second general equation of motion is $v^2 = u^2 + 2as$ .... B

That is, $v^2 = u^2 + 2as_a$

Here s_a is the corresponding value for the acceleration.

The contribution to the $(final\ velocity)^2$ due to the initial velocity is u^2 which does not change.

So, Contribution to v^2 due to initial velocity = u^2 ...... B1

Contribution to the $(final\ velocity)^2$ due to acceleration is,

$$2as_a = 2\frac{v_a}{t}s_a$$

$$= 2v_a\frac{s_a}{t}$$

$$= 2v_a v_a$$

$$= 2v^2_a \quad .... B2$$

Here s_a is the distance contribution and v_a is the velocity contribution due to the acceleration.

So, v^2 = B1 + B2

$$v^2 = u^2 + 2v^2_a \quad .... E$$

That is,

Square of the velocity (v^2) = Contribution from the initial velocity + Contribution from the acceleration

3. The third general equation of motion is $s = ut + \frac{1}{2}at^2$ C

The contribution to the distance s due to the initial velocity u is

$$initial\ velocity \times time = ut$$

Contribution to the distance s due to initial velocity = s_u C1

Here s_u is the contribution of the initial velocity to the distance.

Contribution to the distance s due to the acceleration is

$$\frac{1}{2}at^2 = \frac{1}{2}\frac{v_a}{t}t^2$$
$$= \frac{1}{2}v_a t$$
$$= (\frac{v_a + 0}{2})t$$
$$= \frac{(final\ velocity + initial\ velocity)}{2}t$$
$$= v_{a-aver}\ t$$

v_{a-aver} is the average value* of v_a.

$$= s_a \quad \text{.................... C2}$$

Distance travelled = C1 + C2

$$s = s_u + s_a \quad \text{.................... F}$$

Distance travelled = Contribution from the initial velocity + Contribution from the acceleration

Thus, the general equations of motion (A, B and C) in the new forms are –

$$v = u + v_a \quad \text{............ D}$$

$$v^2 = u^2 + 2v^2{}_a \quad \text{....... E}$$

$$s = s_u + s_a \quad \text{............ F}$$

Here v_a, and s_a are the values related to the acceleration a and s_u is the value related to the initial velocity u.

The initial velocity and the acceleration do not interfere each other and have separate and independent contributions to the final velocity. Equations D and E needs only little clarification since, the velocity and the (velocity)² components of the initial velocity and the acceleration are added to get the final the velocity and the (velocity)².

In equation F, s is the total distance, s_u is the distance corresponding to the initial velocity and s_a is the distance corresponding to the acceleration.

It should be noted that there are two types of distances associated with the motion of an object. They are the measured distance and the actual distance travelled. Actual distance travelled by an object and the measured distance travelled by the object are different due to earth's motion around the sun.

This is true for a free-falling object and for linear motion on earth.

Three cases of an object's motion are shown here using the equation F. In these cases, the earth moves with a constant speed u to the right (in the +X direction) and the time of travel of the object is t and the acceleration for the object for the linear motions on earth is a. The object starts from a point A and ends at point D. During this time, the earth moves from A to B.

Case 1. Free fall of the object (see Figure-1).

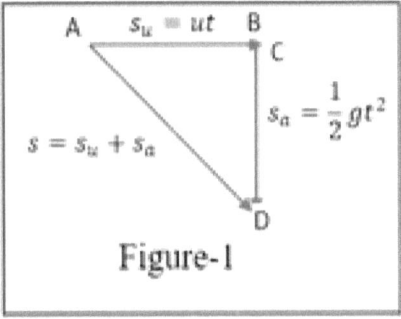

Figure-1

Case 2. Object has a linear motion with an initial velocity in the direction of the motion of the earth (see Figure-2).

```
A                           B
─────────────────────────────▶────▶
                            C    D
```

$AB = s_u = ut$, $CD = s_a = \frac{1}{2}at^2$

$AD = AB + CD = s = s_u + s_a$

Figure-2

Case 3. Object has a linear motion with an initial velocity in the direction opposite to the motion of the earth (see Figure-3).

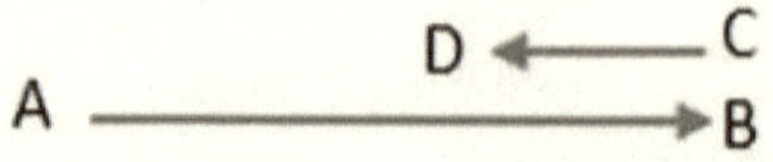

```
                    D ◀──── C
A ──────────────────────▶ B
```

$AB = s_u = ut$, $CD = s_a = \frac{1}{2}at^2$

$AD = AB + CD = s = s_u + s_a$

Figure-3

* Concept of average was introduced by a Belgian, founder of statistics, Adolphe Quetelet (1796–1874)

PAPER 23.
SPACE, TIME, AND DISTANCE RELATIONSHIP

The measurement of separation (non-spatial) between two events is needed in everyday life.

Time is taken as the non-spatial measurement of separation between two events.

Time is measured only in relation to some event in space.

An event occurs in space. The event needs space. Therefore, without space, time has no meaning.

It follows that space and events (multiple) are required for time.

An event is a disturbance in space. That is, the event is a spatial disturbance.

When a reference in time is made yesterday, tomorrow etc.), it is referred to an event or events (physical disturbance/disturbances).

When time interval is considered, it is the interval between two events (spatial disturbances).

This means time in an undisturbed space has no relevance. Distance has no meaning without space.

With this information, time can be defined as follows –

Time is a measurement of non-spatial separation between spatial disturbances.

Similarly,

Distance is a measurement of spatial separation between spatial points.

The definition of time given here will put to rest many questions about time.

1. The question about the beginning and end of absolute time does not arise since time is the measure of separation between events.

2. Since time is the non-spatial separation between events, the question the time can be reversed in direction does not arise.

Event **X1** --Event **X2**
Time (non-spatial separation between the events)

Consider two events X1 and X2. One can say the event X2 occurred after the event X1 or the event X1 occurred before the event X2.

The sequence of occurrence before/after does not mean time travels backward/forward.

3. A non-spatial separation between two events is a scalar quantity

4. A non-spatial separation between events is not a physical entity. Space is physically conceivable (capable of being imagined or grasped mentally), and time is not physically conceivable. That is, it cannot be detected by human senses or imagination (mental exercise).

   An object can be moved from one location in space to another location in space. Time cannot be moved from one place to another.

5. Time does not have an internal mechanism to move.

6. The non-spatial separation (time) of events does not interact with any objects.

7. Time is not a physical quantity. That is, time is not a tangible quantity. Time cannot be detected by physical senses.

8. Time is a physical concept resulting from physical disturbances.

9. In certain cases, two particles are linked even if they are separated by great distance. In this case a change in one particle affects the other. This is called quantum entanglement.

10. Quantum entanglement is considered to be a phenomenon that links space and time. Time arises from physical disturbances. So, Time and

events (physical disturbance) are entangled. Space and distance are entangled.

$$d = \sqrt{(x_2 - x_1)^2 + (y_2 - y_1)^2 + (z_2 - z_1)^2}$$

PAPER 24.

THE TWO STATES OF MATTER – THE STATE OF REST AND THE STATE OF UNIFORM MOTION IN A STRAIGHT LINE.

Newton's first law of motion is dealing with two states of a body – a body at rest and a body in a uniform motion in a straight line.

Do these states of a body are naturally occurring phenomena?

All bodies are made of matter. Matter is made up of energy. Energy packed in a small volume is matter. Energy is dynamic not static. So, motion is an inherent property of all bodies. **That is a body at rest is only a concept of a state of a body.**

Now consider a body in a uniform motion in a straight line. This means the body is covering equal distances in any equal interval of time. How does a body is set into motion? A body is set into motion by a kick (a push or a pull). This is true for all four fundamental forces – gravitation, electromagnetic, strong nuclear and weak nuclear forces.

Gravitational force is a force of attraction (a pull).

Electromagnetic forces are electric forces and magnetic forces. Electric forces are electric force of attraction (a pull) between unlike charges and electric force of repulsion (a push) between like charges. In the same way, magnetic repulsion (a push) and magnetic attraction (a pull).

Strong nuclear force is a force of attraction (a pull) between nucleons inside a nucleus.

Weak nuclear force is the weak interaction among subatomic particles. This is pushing and pulling particles.

The force applied to a body gives an acceleration to the body. In the absence of any other force, this body retains this acceleration – a non-uniform motion. If there is a dissipating force, the body slows down. That is, a force will not produce a uniform motion in a straight line. In addition, according to Newton's third law, for every force there is an equal and opposite reaction. **So, a body with a uniform motion in a straight line is another concept of a state of a body like a body at rest.**

PAPER 25.

MODIFIED STATEMENT OF NEWTON'S THIRD LAW OF MOTION.

Newton's third law of motion states that for every action, there is an equal and opposite reaction.

A couple of examples of action and reaction in everyday life are the followings.

ACTION	REACTION
Centripetal force of a body moving around a circle	Centrifugal force
Weight of a body resting on a table	Reaction on the body by the table

Sometimes the reaction force is called a pseudo force or fictious force since as the action stops, the reaction disappears. Pseudo means not genuine or not real. In fact, it is not the fact that the reaction is there only when the action is there is the question. One can see the physical effects of the reaction makes the reaction real like in centrifuges.

By the current knowledge of the physical universe, the exact mathematical formula of the action can be derived. The formula of the reaction cannot be derived by the available knowledge of the universe. The reaction is a natural reaction. By Newton's third law, one assumes the same equation for the action and its reaction.

For every action, there is an equal and opposite natural reaction.

Where does this reaction come from? Or what is the source of this reaction? This is nature's response to an action.

In the phrase, "Equal and opposite reaction". the word "Opposite" is fully explainable. This means that the direction of the reaction is opposite to the direction of the action. No further explanation or questions are applicable.

What does "Equal" mean? More clarification is needed for the word "Equal". It is obvious that the magnitude of the reaction is equal to the magnitude of the action. What about the format of the reaction? Is the format of the reaction is same as the format of the action? For example, the format of the centrifugal force, which is the reaction of the centripetal force, is the same (that is $m\frac{v^2}{r}$?). Since one does not have sufficient knowledge regarding the universe, centrifugal force is not derived. Can the format of the centrifugal force be $\frac{m}{r}v^2$? – not just rearranging the terms. Or a different format which yields the same magnitude as $m\frac{v^2}{r}$?. These same questions can be raised for any type of reaction.

Reactions referred to in Newton's third law are the mirror reflections of respective actions. Consider an action mirror (an action recorder) with its plane perpendicular to the direction of action (the line of action). The reflection of this action on this mirror is the reaction of this action.

To summarize, there are different actions and their mirror reflections (reactions) in nature. These are action-reaction pairs. In the action-reaction pair, reaction is the reflective complement (counterpart) of the action. Reaction is the reflection of an action in nature. These forces are real and have observable physical effects.

The reaction is reflective meaning reaction appears as soon as the action occurs. The reaction is complement (complementary) means the action will be complete with its reaction.

Thus, Newton's third law of motion can be restated as –

Every action has its reflective complement, the reaction, which is equal and opposite to the action.

PAPER 26.

STRONG FORCE OF REPULSION IN SUPERCONDUCTORS.

Similar electric charges repel. A proton repels another proton, and an electron repels another electron. Under certain conditions, similar charges can stay close together/combined. Protons can stay close together/combined, and electrons can stay close together/combined.

Protons can stay together like in atomic nucleus and electrons (as pairs) stay close together like in super conductors.

Protons stay together in nucleus is explained (in Paper 5.2) by sharing the mass defects between the nucleons.

When positively charged particles (protons) stay close together (or combined together), there will be a strong attractive force (nuclear force) – **protons-neutrons strong force**.

When negatively charged particles (electrons) stay close together (or combined), they produce a repulsive force (**electrons strong force**.) in the immediate neighborhood. **This is a new type of force.**

Both these forces (protons-neutrons strong force and electrons strong force) are very strong and short ranged.

It is observed that the electrons in super conductors in a pair (2 electrons) – not more than two. This shows that as soon as the two electrons join together, they form an opposing force in their immediate neighborhood.

Mass is a packet of energy. Energy is dynamic not static. So, energy is in constant motion. Circular motion is a characteristic of stable system from micro to macro systems. There is circular motion in atoms to planetary system. So, the electrons joined here have a circulatory motion. The centrifugal force of this circular motion is the opposing force of these electrons to other particles in their immediate neighborhood.

It is unlikely that the two opposing electrons stay close together as two independent electrons. It is highly possible that these two electrons combine to one particle, an electron double.

In super conductors, electrons are paired up (stays close together/combined). They produce a strong repulsive force in its neighborhood. This avoids any particles near these paired electrons. So, these electrons move without any contact with other particles in a super conductor. This makes the electrical resistance in a superconductor zero.

This repulsive force produces a pressure outward in the medium away from the moving electron doubles.

In normal electric flow of charges, the electrons collide with neighboring particles on their way resulting in the heating of the conductor. As the electric current increases, the collisions increase, and the conductor heats up more.

Likewise in superconductors, the electron doubles have opposing force on the neighboring particles and there will be a pressure increase in the conductor. As more electron doubles flows, electric current increases, the pressure on the neighboring particles increases.

The increase in pressure by the electron doubles in the medium away from the moving electron doubles can be detected.

PAPER 27.
MATERIAL ADDED TO PAPER 5.2

Following material is added in October 2022 for the Paper 5.2 written in 2018.

One of the main arguments by X. D. Dongfang in an interesting article "*The End of Yukawa Meson Theory of Nuclear Forces**" is that "there has never been a meson in the nucleus."

---

*MATHEMATICS & NATURE Mathematics, Physics, Mechanics & Astronomy, 2021 Vol. 1 No. 1: 010 DOI: 10.13140/RG.2.2.21931.31529

PAPER 28.
ACTION, REACTION AND CENTRIGUGAL FORCE

According to Newton's third law, every action has an equal and opposite reaction. One sees the action-reaction pair in numerous cases in everyday life. Some examples are walking, swimming, recoil of a gun, holding a pencil.

All these involves mass. In all these, the reaction is the collective reaction of the particles which are impacted by its cause, the action.

The centrifugal force exerted on a particle moving around a circle is different. Figure-1 shows the centripetal and centrifugal forces on a particle moving around a circle with a constant speed v.

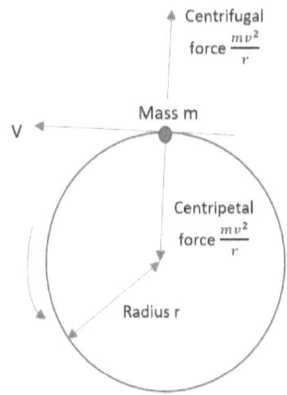

Figure-1

The centripetal and centrifugal forces are the same. The centrifugal force is not coming from any material particle. **The centrifugal force is the reaction of space against the centripetal force.**

So, it may seem that the centrifugal force is an exception where the reaction is coming from space. But nature always applies its rules and regulations equally everywhere. That is even if there is material or not, involved in an action, the space produces a reaction, and this reaction is propagated to the particles present in that space.

When the centripetal acceleration changes (increase/decrease), the centrifugal force adjusts accordingly (increase/decrease) so as to make the centrifugal force equal to the centripetal force. The mass remains the same. **So, strictly speaking, the centrifugal acceleration is equal to the centripetal acceleration but opposite in direction**.

This is important. Consider an example where an entity X remains the same before and after a certain process. Let this X consist of the following 5 items.

Item-a, item-b, item-c, item-d, and item-e

If items-a, item-b and item-c remain the same (constant) before and after the process, item-d increases, and item-e decreases by the same amount during the process as to make X the same before and after the process. A statement saying that X is the same before and after the process tells only a part of the story and does not give an insight to the process. For a better analysis, the statement should be that item-a, item-b and item-c remain the same before and after the process and item-d increases and item-e decreases by the same amount.

PAPER 29.
REFERENCE FRAMES AND FIRST POSTULATE OF SPECIAL THEORY OF RELATIVITY

What is a **frame of reference**?

A reference frame is a group of points in space, where there is no relative motion among the points. A reference frame is also referred to as frame of reference or just as a frame. An **observational frame of reference**, often referred to as a physical frame of reference with physical reference points fixed on observational objects. The two types of reference frames are inertial and non-inertial.

What is an inertial reference frame?

If a reference frame has no acceleration there is no force acting on the body in this frame. That is, the motion of a body or systems of bodies is determined entirely by its inertia. Such a reference frame is called an **inertial frame**. An inertial frame is a frame where Newton's laws of motion are true. A small portion of earth's orbital motion around the sun is nearly an inertial frame.

What is a non-inertial reference frame?

A non-inertial frame is a frame which has acceleration. They have motions that change the speed or direction or both. Newton's laws are not valid in a non-inertial frame.

Einstein published the special theory of relativity in 1905. The word special here means that this theory is restricted to observers in an inertial frame. The general theory of relativity which was introduced by Einstein in 1916, is for observers in any state of uniform motion including relative acceleration.

The **first postulate of relativity** states that the laws of physics are the same in all inertial frames. This also means that the laws of physics can never be expressed in terms of the speed of the inertial reference frame where it is valid. The law of physics is equally valid in inertial frames moving with uniform velocities 5 meters/second and 5000 meters per second.

Two immediate and direct implications of this first postulate are the following.

If one does an experiment within a reference frame without any observation to the outside world,

1. the motion of the frame cannot be detected. and

2. the velocity of the reference frame cannot be measured.

Examine the first implication. Can one find out (not by any measurement) somehow an inertial frame of reference is moving with the help of any experiment performed within the frame? Consider the following steps.

1. Laws of physics are valid in all inertial frames.
2. Inertial frames are those frames which are at rest or frames with uniform motion.
3. Motion is an inherent property of all bodies and hence all frames of reference. That is, there is no physical frame which is at rest. Also, according to the special theory of relativity, there is no resting frame.
4. Suppose one tests one of Newton's laws of motion (a law of physics) in a reference frame and the law is found to be true. Since there is no reference frame which is at rest, the person determines that the reference frame is an inertial frame with uniform motion.

That is, one can do an experiment within a reference frame and find out the nature of the reference frame.

Initially the first postulate of special relativity was applied to electrodynamics and optics. Later this postulate was applied to all physical laws. The question is whether this generalization is valid? Assuming it is valid one can further proceed as fallows

The laws of physics are derived in an inertial frame (frame with uniform motion) without considering the motion of the frame. So, these laws are valid in any other inertial frames without regard to the speed of the frame.

Some results of the special theory of relativity seem to contradict our common sense. According to Albert Einstein " Common sense is a collection of prejudices acquired by the age eighteen". Two points worth mentioning about this. First some of these results may remain contradictory to our common sense until a sensible explanation is given. Second, common sense is a preconceived opinion based on reason and actual experience. Common sense is the human sense acquired by the surrounding natural events (physical disturbances see Paper 23), their consequences and their sequence of occurrences like true and false, cause and effect and other effects of the events.

PAPER 30.
ENTANGLED PHOTONS

Entangled photons are those photons which behave same way even if they are separated by far distance.

Space is not uniform/homogeneous.

Its characteristics vary from place to place.

In two nearby physical points, space is rather uniform and more or less with the same characteristics.

Considering the half wave energy packets explained in Paper 8, two nearby energy packets are near spaces and inherit rather uniform space characteristics. That is, two energy packets are formed at the same time in the same or nearby space with same or almost same characteristics.

So, two photons produced close by behave the same way.

That is two nearest photons are entangled more compared to two photons created at far away physical points.

Space is not uniform. Its characteristics vary from place to place. In two nearby physical points, the space is uniform and more or less with same characteristics. Considering the half wave energy packets explained in Paper 8, two nearby energy packets are near spaces and inherit rather uniform space characteristics. So, two photons produced close by behave the same way. That is two close by photons are entangled more compared to two photons created at far away physical points.

PAPER 31.
RULES OF THE BASIC LAWS OF PHYSICS

There are some basic laws of physics like Newton's three laws of motion, law of gravitation.

Rules governing these basic laws of physics are

1. The basic laws of physics are simple and sensible to human beings.

2. The basic laws of physics are unique.

3. one basic laws of physics should not contradict, interact or overlap another basic law.

4. Any one physical phenomenon can be explained by only one of the basic laws of physics.

5. For a given set of physical quantities, there is only one basic law of physics.

6. As of now there is no known law between two type of physical quantities (like mass and charge).

   When two charges of different masses interact, the larger mass results in less acceleration. This is not because the interaction between the two charges is different but because of Newton's second law of motion.

7. If interactions due to two basic laws of physics exist in a close space, there will be no interference between the two interactions.

8. If a basic law of physics involves the interaction between two physical quantities, it will be by the product of two items (the product law - refer Paper 1. Paper 3, Paper 4.1 and Paper 4.2).

PAPER 32.
FLOATING OR SINKING BODY IN A FLUID

Archimedes' principle (246 BC from Greek city of ancient Sicily) is considered as a basic physics concept, a fundamental concept in fluid mechanics.

Floating/sinking body in a fluid and its apparent loss of weight is another example of inseparable and ever together action and reaction pair of forces. There is an action of the body on the fluid and the fluid has a reaction on the body.

A sinking or a floating object in a fluid has an action on the fluid. This action is to displace the fluid. Which physical property of the displaced fluid is relevant here. The weight of the displaced liquid is that matters here.

This action is displacing a liquid of certain weight acting down. So, there is a reaction of the fluid which is equal (the weight of the displaced fluid) and opposite to this action. So, the floating or sinking body feels an apparent loss of weight.

The net weight of the body = weight of the body
— weight of the fluid displaced.

If the body floats on the surface of a fluid, its total weight is compensated by the weight of the fluid it displaced.

If the body keeps sinking in the fluid, at any height from the surface of the fluid, its partial weight (weight loss) is compensated by the weight of the fluid displaced at that level. The weight of equal volume of fluid at different heights from the bottom are different – least on the surface and maximum at the bottom. That is, its loss of weight is the weight of the displaced fluid.

If the body reaches the bottom of the fluid, its loss of weight is the weight of the fluid it displaced at that level. The effective weight (net weight) of the body is compensated for by the reaction of the bottom of the fluid.

This is like a body resting on a table. The weight of the book acting downwards is compensated by a reaction of the table acting on the book upwards.

PAPER 33.
PARTICLE AND SPACE SIGNALS

There are two types of signals in nature. They are particle signals and space signals. Particle signals contain physical particles. Space signals do not contain any physical particles. A ray of light is a particle signal. Newton's actions at a distance - Gravitational forces (signals) – mass effect (Paper 3), Electric and magnetic forces (signals) – charge effect (Paper 4.1 and 4.2) are space signals. Phenomenon where paired particles instantaneously correlate their states regardless of the distance separating them is another example of space signals.

Space signals start from physical objects. Space signals can affect only another object of same type. For example, signals from mass (mass effect) affect only mass. Space signals can cause attraction of objects, can cause repulsion of objects or can cause a physical change in another object – of same type.

Space signals are two types – **distance dependent signal** (causing action at a distance) and **distance independent signal** (causing action at a distance). Mass effects (Paper 3) and charge effects (Paper 4.1 and 4.2) are due to distance dependent signals. That means at large distances, these effects will be extremely small. Phenomenon where paired particles instantaneously correlate their states is due to distance independent signals.

These two actions from the space signals should be termed appropriately. Both are resulting from particles. One is distance dependent and other is distance independent. Distance dependent action can be termed as **action at a distance**. Distance independent action can be termed as **action at a point**.

The product law introduced in Paper 1 is applied to mass effect in Paper 3 and applied to charge effect in Paper 4.1 and in 4.2. This product law should be applicable to action at a point also. That is, if two particles A and B are involved in an action at a point, A has an action at a point on B and B has an action at a point on A, these two actions interact. This interaction is the product of individual actions at a points at A and B.

In the derivation of mass energy equation $E = mc^2$, laws of particle world are considered. In a particle world, if light signal is the maximum attainable speed, the assumption that Speed of light (c) as the universal speed limit, is justifiable.

NOTES – PAGE 1

NOTES - PAGE 2

NOTES – PAGE3

NOTES - PAGE 4

NOTES – PAGE 5

NOTES – PAGE 6

NOTES – PAGE 7

NOTES - PAGES

www.ingramcontent.com/pod-product-compliance
Lightning Source LLC
Chambersburg PA
CBHW030947240526
45463CB00016B/2040